Recurrence and Topology

Recurrence and Topology

John M. Alongi
Gail S. Nelson

Graduate Studies in Mathematics
Volume 85

American Mathematical Society
Providence, Rhode Island

2000 *Mathematics Subject Classification.* Primary 37-01, 37B20, 37B25, 37B35, 54H20; Secondary 37C10, 37C15, 37C25, 37C27, 37C50, 37C70, 34D45.

For additional information and updates on this book, visit
www.ams.org/bookpages/gsm-85

Library of Congress Cataloging-in-Publication Data

Alongi, John M., 1971–
Recurrence and topology / John M. Alongi, Gail S. Nelson.
p. cm. — (Graduate studies in mathematics, ISSN 1065-7339 ; v. 85)
Includes bibliographical references and index.
ISBN-13: 978-0-8218-4234-8 (alk. paper)
ISBN-10: 0-8218-4234-X (alk. paper)
1. Recurrent sequences (Mathematics) 2. Point mappings (Mathematics) 3. Topology. I. Nelson, Gail Susan. II. Title.

QA9.6.A46 2007
515′.24–dc22 2007060754

---

10 9 8 7 6 5 4 3 2 1 12 11 10 09 08 07

Contents

Foreword

During the early nineteenth century the French mathematician Simeon Denis Poisson (1781–1840) observed through his study of celestial mechanics that the solutions of a differential equation which exhibit recurrent behavior more general than periodicity play a central role in determining the disposition of all solutions of the system. Since the time of Poisson, mathematicians have sought precisely what it means for a solution of a differential equation to be recurrent.

Recurrence and Topology develops increasingly more general topological modes of recurrence for dynamical systems beginning with fixed points and concluding with chain recurrent points. For each type of recurrence we provide detailed examples arising from explicit systems of differential equations; we establish the general topological properties of the set of recurrent points; and we investigate the possibility of partitioning the set of recurrent points into subsets which are dynamically irreducible. Furthermore, we consider how test functions such as invariant functions, potential functions and Lyapunov functions describe the structures of sets of recurrent points. The text concludes with a statement, proof, and interpretation of the Fundamental Theorem of Dynamical Systems due to Charles Conley (1933–1984).

Recurrence and Topology has a deliberately narrow focus. We treat flows (continuous dynamical systems) rather than maps (discrete dynamical systems) for three reasons. First, flows arise directly and naturally from the differential equations that motivate the subject. Second, the connectedness of orbits for flows sets the stage for a rich interplay between recurrence and topology. Third, chains for flows rely on both distance and time, while chains for maps rely only on distance.

Recurrence and Topology does not treat results, such as the Poincaré-Bendixson Theorem, which are particular to low dimension, nor do we discuss recurrence in the context of measure theory nor recurrence arising from hyperbolicity. However, we hope there is a niche for this text in seminar courses, independent studies, and as a supplement to comprehensive works such as Katok and Hasselblatt [**26**] and Robinson [**37**] which develop many threads within the subject simultaneously. *Recurrence and Topology* examines a single thread and aims to present it in a clear, complete and coherent manner.

While the objective of *Recurrence and Topology* is to integrate existing knowledge, we offer detailed original proofs of a few folklore theorems (such as the fact that a flow's chain components are exactly the connected components of the flow's chain recurrent set). We present competing definitions of certain terms (such as what it means for a set to be topologically transitive with respect to a flow) and argue for the definitions we adopt.

With the work of Stephen Smale (1930–) during the 1960s and the genesis of chaos as a scientific paradigm during the mid-1970s, the field of dynamical systems enjoyed a renaissance. Today it seems appropriate to reflect on advances in the field, to place them in context, and to make them accessible to a broader mathematical audience. We hope that *Recurrence and Topology* is a step in that direction.

Recurrence and Topology is appropriate for mathematics graduate students, though a well-prepared undergraduate might read most of the text with great benefit. We presume differential equations, undergraduate analysis and general topology as background. The texts by Boyce and DiPrima [**10**], Rudin [**38**] and Munkres [**32**] provide sufficient preparation. Certain sections also require advanced linear algebra and complex variables. After referring to manifolds in Section 1.1 for motivation, we do not require the geometry of manifolds until Sections 4.5 and 4.6. Milnor [**31**], Guillemin and Pollack [**17**], and Hicks [**20**] are good references for the relevant material. Regarding our exposition, we emphasize clarity over brevity and apologize to readers who find our presentation excessively detailed.

John Alongi thanks John Franks, Clark Robinson, Bob Welland, Don Saari, Jeff Xia and Keith Burns for teaching him about dynamical systems, topology and differential equations. Thanks to the Department of Mathematics and Computer Science at Pomona College for providing financial support from its Spears account during the 2003–2004 and 2004–2005 academic years. Most of all, thanks to Michael and Joan Alongi, David, Tina and Kristine Alongi, Ryan Beveridge and Tricia Hansen, Alissa Crans, Allison Engel, and Connie and Jim Pemble.

Gail Nelson thanks Steve Kennedy, Sam Patterson, and John Alongi for inviting her to join their research seminar, giving her the opportunity to learn more about dynamical systems. Thanks also to the rest of the members of the Department of Mathematics at Carleton College for their support of this project. Many thanks to her family, Rik, Julie, and Lauren Getty, Doug, Teresa, Thomas, and Anna Bearrood, and especially Edward Nelson, who has always been her role model.

Both authors thank Joseph Borzellino, Karen Brucks, Ezra Buchla, Richard Elderkin, Francesca Fairbrother, John Franks, Barbara Jenkins, Stephen Kennedy, Laurene Lee, Sam Patterson, Ami Radunskaya, Clark Robinson, Anthony W. Sevold and Kathy Sheldon for their assistance in this endeavor. In particular, Ezra Buchla and Anthony W. Sevold rendered the figures in the text. Laurene Lee offered many important suggestions while reading a draft of the text. Both authors also thank the anonymous reviewers of the text as well as the editors and staff of the American Mathematical Society for their thoughtful suggestions. Ed Dunne, Cristin Zannella and Luann Cole of the AMS deserve special recognition in this regard.

Please direct comments and corrections to `jalongi@northwestern.edu` or `gnelson@carleton.edu`.

John M. Alongi
Chicago, IL

Gail S. Nelson
Northfield, MN

Chapter 1

Flows

1.1. Flows and Orbits

Let M be a compact smooth manifold without boundary, and let f be a smooth vector field on M. Consider the differential equation

$$\dot{x} = f(x)$$

where $\dot{x}$ denotes the derivative of a function x with respect to a single independent variable. We usually denote the independent variable of x by t since this variable often signifies time when the differential equation models a natural process.

The existence and uniqueness theorem for ordinary differential equations guarantees that there exists a unique smooth function $\phi : \mathbb{R} \times M \to M$ such that

(i) $\phi(t, \cdot) : M \to M$ is a diffeomorphism for each $t \in \mathbb{R}$;

(ii) $\phi(t + s, x) = \phi(s, \phi(t, x))$ for all $x \in M$ and all $s, t \in \mathbb{R}$; and

(iii) for each $x \in M$,

$$\left.\frac{d}{dt}\right|_{t=0} \phi(t, x) = f(x).$$

Chapter 5 of Spivak [**43**] establishes the existence of a unique function ϕ satisfying properties (i)–(iii). Pages 371–379 of Lang [**28**] contain a proof that ϕ is smooth.

For each $x_0 \in M$ the function $\phi(\cdot, x_0) : \mathbb{R} \to M$ is called a **solution** of the differential equation, and its image is called an **integral curve** for the vector field f. The value $\phi(t, x_0)$ is the point obtained by following the solution of $\dot{x} = f(x)$ satisfying the initial condition $x(0) = x_0$ for time t.

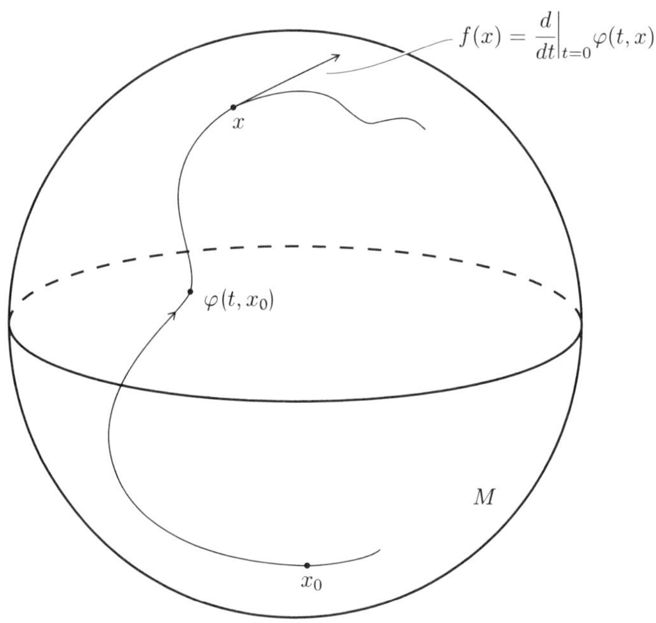

Figure 1.1. A solution of a differential equation on a manifold

The vector $f(\phi(t, x_0))$ is the velocity vector at time t of the solution $\phi(\cdot, x_0)$. See Figure 1.1. The function ϕ is a single mathematical object that captures the solutions of $\dot{x} = f(x)$ with respect to all initial conditions at once.

For much of our discussion it suffices to consider two important cases for the manifold M:

- M is the circle S^1.
- M is a compact surface without boundary such as the sphere S^2 or the torus $\mathbb{T}^2$.

However, we shall also consider examples in which a differential equation $\dot{x} = f(x)$ on M determines a function $\phi : \mathbb{R} \times M \to M$ satisfying properties (i)–(iii) even though M is either not compact or has a nonempty boundary. For our purposes there are three important situations in which this occurs:

- M is a Euclidean space $\mathbb{R}^n$.
- M is a compact interval $I \subset \mathbb{R}$.
- M is the closed unit disk $\mathbb{D}$ in the complex plane $\mathbb{C}$.

To study topological features of functions ϕ arising from differential equations in this way, we extract their elementary topological properties from (i)–(iii) by generalizing from a manifold to a topological space and by relaxing the differentiability conditions to continuity conditions.

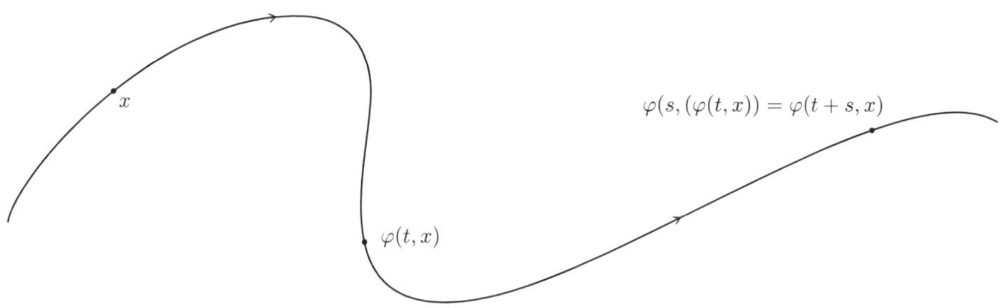

Figure 1.2. The group property of flows

Definition 1.1.1. A **flow** (or **complete flow** or **continuous dynamical system**) on a topological space X is a continuous function $\phi : \mathbb{R} \times X \to X$ such that

(i) $\phi(t, \cdot) : X \to X$ is a homeomorphism for each $t \in \mathbb{R}$, and

(ii) $\phi(t + s, x) = \phi(s, \phi(t, x))$ for all $x \in X$ and all $s, t \in \mathbb{R}$.

Property (ii) is the **group property of flows**. See Figure 1.2. The topological space X is the **phase space** of the flow ϕ.

Proposition 1.1.2 follows directly from Definition 1.1.1 and reflects the fact that a flow is a continuous action of the topological group $(\mathbb{R}, +)$ on the flow's phase space. See Exercise 5. This connection to group actions explains why we refer to Definition 1.1.1 (ii) as the group property of flows.

Proposition 1.1.2. *If ϕ^t is a flow on a topological space X, then*

(i) *ϕ^0 is the identity map on X, and*

(ii) *for each real number t the map ϕ^t is invertible, and $(\phi^t)^{-1} = \phi^{-t}$.*

Proof. (i) By the group property,

$$\phi^0(x) = \phi^0(\phi^0(x))$$

for all $x \in X$. According to Definition 1.1.1(i) the map $\phi^0 : X \to X$ is injective. Consequently, $\phi^0(x) = x$ for all $x \in X$. Therefore, ϕ^0 is the identity map on X.

(ii) According to Definition 1.1.1(i) the map ϕ^t is invertible for each $t \in \mathbb{R}$. By the group property and part (i),

$$\phi^t(\phi^{-t}(x)) = \phi^{-t}(\phi^t(x)) = \phi^0(x) = x$$

for each $t \in \mathbb{R}$ and all $x \in X$. Therefore,

$$(\phi^t)^{-1} = \phi^{-t}$$

for each $t \in \mathbb{R}$. □

For each real number t, a flow $\phi : \mathbb{R} \times X \to X$ defines a function $\phi(t, \cdot) : X \to X$. A function from a set to itself is called a map.

Definition 1.1.3. Let X be a set. A **map** (or **discrete dynamical system**) is a function $f : X \to X$.

Customarily, we write $\phi^t(x)$ for $\phi(t, x)$ and denote a flow $\phi : \mathbb{R} \times X \to X$ by $\phi^t : X \to X$. In this notation the group property appears as

$$\phi^{t+s}(x) = \phi^s(\phi^t(x))$$

for all $x \in X$ and all $s, t \in \mathbb{R}$. However, the notation ϕ^t may also signify the map $\phi(t, \cdot)$ for a fixed real number t. We endeavor to make the meaning of the notation ϕ^t clear from its context.

The map ϕ^1 is useful when relating the behavior of a flow to the behavior of a map.

Definition 1.1.4. If ϕ^t is a flow on a topological space X, then the function $\phi^1 : X \to X$ is called the **time-one map** of ϕ^t.

In general, obtaining an explicit expression for the flow corresponding to a differential equation involves integrating the associated vector field. Such integrals are not usually elementary functions. See Kaplansky [**25**]. Thus, finding an explicit expression for a flow is often impossible. Nonetheless, there are important examples of flows for which explicit formulas are available.

Example 1.1.5. Let $c \in \mathbb{R}^n$. Consider the **constant differential equation**

$$\dot{x} = c$$

on $\mathbb{R}^n$.

The flow $\phi^t : \mathbb{R}^n \to \mathbb{R}^n$ associated to this differential equation is

$$\phi^t(x) = x + ct.$$

The differential equation $\dot{x} = c$ defines a flow even though $\mathbb{R}^n$ is not compact.

Example 1.1.6. Let A be an $n \times n$ matrix with real entries. Consider the **linear differential equation**

$$\dot{x} = Ax$$

on $\mathbb{R}^n$. The flow $\phi^t : \mathbb{R}^n \to \mathbb{R}^n$ associated to this differential equation is

$$\phi^t(x) = e^{At}x$$

where the matrix exponential e^{At} is the matrix defined by

$$e^{At} = \sum_{k=0}^{\infty} \frac{(At)^k}{k!}.$$

Chapter 3, §14–§15 of Arnol'd [**4**] develop the properties of this series. In particular, it is permissible to differentiate this series term-by-term to obtain

$$\frac{d}{dt}e^{At} = Ae^{At}.$$

As in Example 1.1.5, the differential equation $\dot{x} = Ax$ defines a flow even though $\mathbb{R}^n$ is not compact.

Example 1.1.7. Consider the **logistic differential equation**

$$\dot{x} = x(1-x)$$

on $[0,1]$. Using separation of variables and integration by partial fractions, the flow $\phi^t : [0,1] \to [0,1]$ corresponding to the logistic equation is

$$\phi^t(x) = \frac{xe^t}{1 + x(e^t - 1)}.$$

Example 1.1.8. Consider the differential equation

$$\dot{\theta} = \sin^2 \theta$$

in the angular coordinate θ on S^1. Using separation of variables, the flow $\phi^t : S^1 \to S^1$ associated to this differential equation is

$$\phi^t(\theta) = \begin{cases} \cot^{-1}\beta + 2n\pi & \text{if } \beta \in I_{2n} \text{ for some } n \in \mathbb{Z}, \\ \cot^{-1}\beta + (2n+1)\pi & \text{if } \beta \in I_{2n+1} \text{ for some } n \in \mathbb{Z}, \\ \theta & \text{if } \theta = n\pi \text{ for some } n \in \mathbb{Z}, \end{cases}$$

where $\beta = \cot\theta - t$, $I_{2n} = (2n\pi, (2n+1)\pi)$ and $I_{2n+1} = ((2n+1)\pi, 2(n+1)\pi)$. We shall refer to this flow as the **arrested rotation**.

Orbits generalize integral curves.

Definition 1.1.9. If ϕ^t is a flow on a topological space X and $x \in X$, then the **orbit** of x with respect to ϕ^t is the set

$$\mathcal{O}(x) = \{\phi^t(x) \mid t \in \mathbb{R}\}.$$

The **forward orbit** of x with respect to ϕ^t is the set

$$\mathcal{O}^+(x) = \{\phi^t(x) \mid t \geq 0\}.$$

The **backward orbit** of x with respect to ϕ^t is the set

$$\mathcal{O}^-(x) = \{\phi^{-t}(x) \mid t \geq 0\}.$$

When it is necessary to explicitly identify the flow ϕ^t we denote the orbit of x, the forward orbit of x and the backward orbit of x by $\mathcal{O}(x, \phi^t)$, $\mathcal{O}^+(x, \phi^t)$ and $\mathcal{O}^-(x, \phi^t)$, respectively.

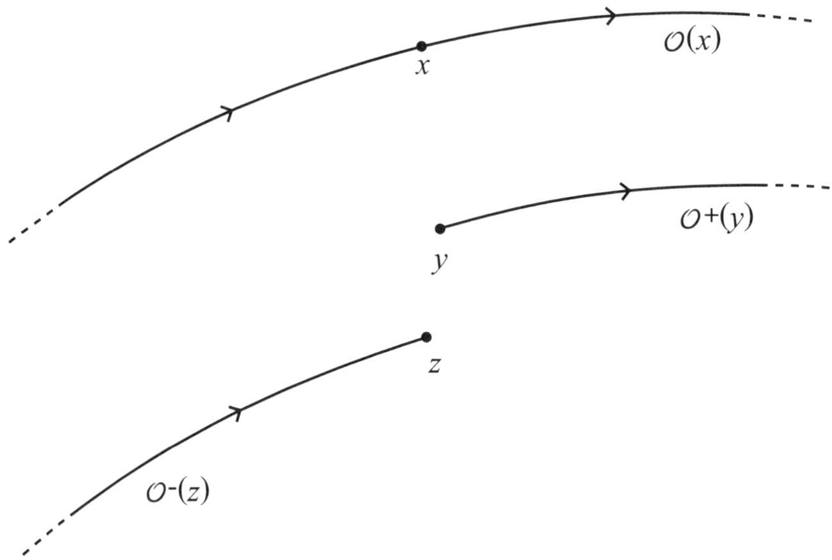

Figure 1.3. An orbit, a forward orbit and a backward orbit

If ϕ^t is a flow on a topological space X and $x \in X$, then the orbit of x is the image of the path $\phi(\cdot, x) : \mathbb{R} \to X$. Thus, every orbit inherits an orientation (a preferred direction) from the positive orientation of $\mathbb{R}$ (the direction of increasing t in $\mathbb{R}$). See Figure 1.3. Appendix A defines the orbit of a point with respect to a map.

Example 1.1.10. Let $c \in \mathbb{R}^n$. Consider the constant differential equation

$$\dot{x} = c$$

on $\mathbb{R}^n$. By Example 1.1.5 the flow ϕ^t associated to this differential equation is

$$\phi^t(x) = x + ct.$$

Let $x \in \mathbb{R}^n$. The orbit of x under ϕ^t is

$$\mathcal{O}(x) = \{\phi^t(x) \mid t \in \mathbb{R}\} = \{x + ct \mid t \in \mathbb{R}\}.$$

If $c \neq 0$, then $\mathcal{O}(x)$ is the line in $\mathbb{R}^n$ through x in the direction of c. See Figure 1.4. If $c = 0$, then the orbit of x is $\{x\}$.

Since orbits are images of paths, orbits, forward orbits and backward orbits are connected.

Proposition 1.1.11. *The orbit, forward orbit and backward orbit of a point with respect to a flow are path-connected spaces; hence, they are connected spaces.*

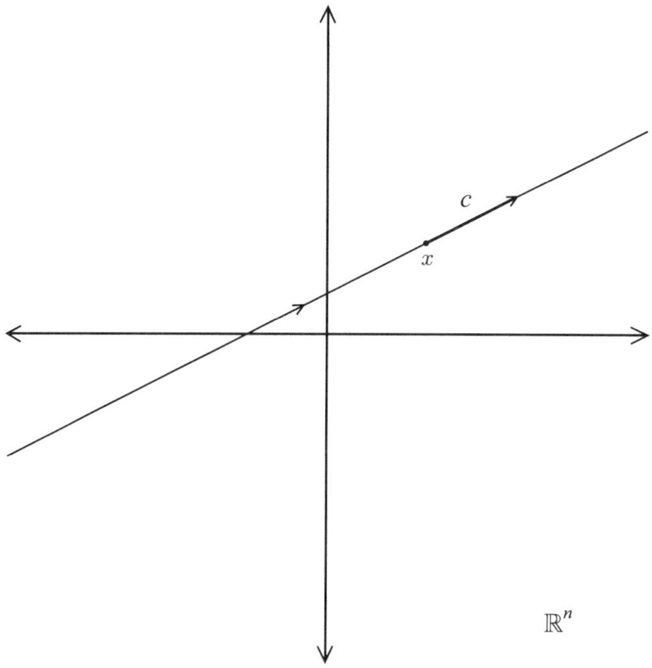

Figure 1.4. An orbit of the flow of a constant differential equation

Proof. Let ϕ^t be a flow on a topological space X. Let $x \in X$. Since $\phi^t(x) : \mathbb{R} \to X$ is continuous as a function of t, its image $\mathcal{O}(x) = \{\phi^t(x) \mid t \in \mathbb{R}\}$ is path-connected. Every path-connected space is connected. Similarly, $\mathcal{O}^+(x)$ and $\mathcal{O}^-(x)$ are path-connected and connected. $\square$

Despite its simplicity, Proposition 1.1.11 identifies a crucial topological difference between flows and maps. An orbit with respect to a flow must be connected, whereas Example A.0.3 demonstrates that an orbit with respect to a map may not be connected.

The existence and uniqueness theorem for ordinary differential equations states that if f is a smooth vector field, then the differential equation $\dot{x} = f(x)$ has a unique solution satisfying the initial condition $x(0) = x_0$. Thus, if two integral curves each contain x_0, then the integral curves are equal. More generally, orbits which intersect actually coincide.

Proposition 1.1.12. *If two orbits have a nonempty intersection, then the orbits are equal.*

Proof. Let ϕ^t be a flow on a topological space X. Let $x, y \in X$. If $\mathcal{O}(x) \cap \mathcal{O}(y)$ is nonempty, then there exists $z \in \mathcal{O}(x) \cap \mathcal{O}(y)$. So, there exist real

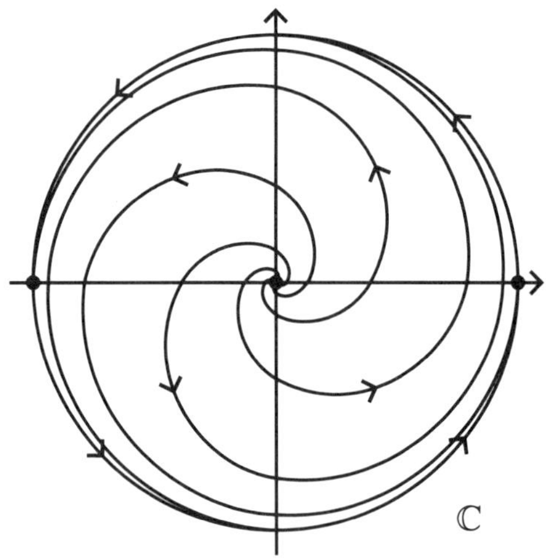

Figure 1.5. A phase portrait

numbers r and s such that $z = \phi^r(x) = \phi^s(y)$. Hence, $x = \phi^{-r}(z)$ by Proposition 1.1.2(ii).

Let $w \in \mathcal{O}(x)$. There exists a real number t such that $w = \phi^t(x)$. By the group property of flows,

$$w = \phi^t(\phi^{-r}(z)) = \phi^t(\phi^{-r}(\phi^s(y))) = \phi^{t-r+s}(y) \in \mathcal{O}(y).$$

Thus, $\mathcal{O}(x) \subseteq \mathcal{O}(y)$. Similarly, $\mathcal{O}(y) \subseteq \mathcal{O}(x)$. Therefore, $\mathcal{O}(x) = \mathcal{O}(y)$. □

A **phase portrait** of a flow is a pictorial representation of the flow's phase space including representative orbits of the flow and arrows indicating the orientations of those orbits with respect to increasing time. Figure 1.5 shows a phase portrait of a flow which will be important throughout the text.

1.2. Topologically Conjugate Flows

In this section and the next we identify two ways in which flows may be topologically similar. The stronger of these two types of similarities is topological conjugacy. To motivate the definition of topologically conjugate flows we examine linear differential equations.

Example 1.2.1. Let A and B be similar $n \times n$ matrices with real entries. Consider the linear differential equation

$$\dot{x} = Ax$$

on $\mathbb{R}^n$. By Example 1.1.6 the flow $\phi^t : \mathbb{R}^n \to \mathbb{R}^n$ associated to this differential equation is

$$\phi^t(x) = e^{At}x,$$

and the flow $\psi^t : \mathbb{R}^n \to \mathbb{R}^n$ corresponding to the differential equation

$$\dot{x} = Bx$$

is

$$\psi^t(x) = e^{Bt}x.$$

Since A and B are similar, there exists an invertible matrix S such that

$$BS = SA.$$

Equivalently,

$$SAS^{-1} = B.$$

Thus,

$$\begin{aligned} Se^{At}S^{-1} &= S\left[\sum_{k=0}^{\infty} \frac{(At)^k}{k!}\right]S^{-1} \\ &= \sum_{k=0}^{\infty} S\frac{(At)^k}{k!}S^{-1} \\ &= \sum_{k=0}^{\infty} \frac{(SAS^{-1}t)^k}{k!} \\ &= \sum_{k=0}^{\infty} \frac{(Bt)^k}{k!} \\ &= e^{Bt}. \end{aligned}$$

Thus, e^{At} and e^{Bt} are similar matrices, and

$$e^{Bt}S = Se^{At}$$

for all real numbers t.

Since S is an invertible matrix, S represents a linear isomorphism $L : \mathbb{R}^n \to \mathbb{R}^n$ such that

$$\psi^t \circ L = L \circ \phi^t$$

for all real numbers t. We say that L conjugates ϕ^t and ψ^t.

A linear isomorphism is a bijection that preserves the structure of a linear space. A bijection that preserves the structure of a topological space is a homeomorphism. In general, we define two flows to be topologically conjugate if there is a homeomorphism that conjugates them.

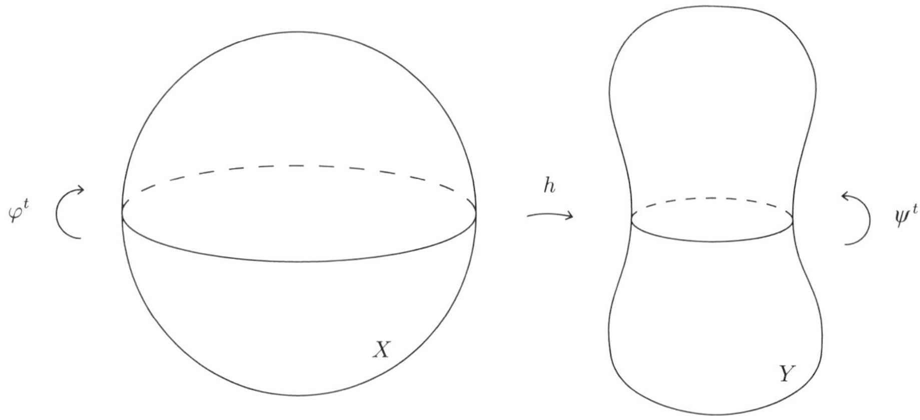

Figure 1.6. Topological conjugacy

Definition 1.2.2. Let X and Y be topological spaces. The flows $\phi^t : X \to X$ and $\psi^t : Y \to Y$ are **topologically conjugate** if there exists a homeomorphism $h : X \to Y$ such that

$$\psi^t \circ h = h \circ \phi^t$$

for all real numbers t. The function h is a **topological conjugacy** from ϕ^t to ψ^t. See Figure 1.6.

Example 1.2.3. If A and B are similar matrices with real entries, then the flows ϕ^t and ψ^t of Example 1.2.1 are topologically conjugate, and the linear isomorphism L of Example 1.2.1 is a topological conjugacy from ϕ^t to ψ^t.

If $h : X \to Y$ is a topological conjugacy from $\phi^t : X \to X$ to $\psi^t : Y \to Y$, then the diagram

$$\begin{array}{ccc} X & \xrightarrow{\phi^t} & X \\ h\downarrow & & \downarrow h \\ Y & \xrightarrow[\psi^t]{} & Y \end{array}$$

commutes for each real number t.

A topological conjugacy maps orbits to orbits. See Exercise 7. Furthermore, a topological conjugacy preserves the time parametrization of each orbit.

1.3. Topologically Equivalent Flows

We may also consider two flows ϕ^t and ψ^t to be topologically similar if there is a homeomorphism mapping the orbits of ϕ^t to the orbits of ψ^t with possibly different time parametrizations while preserving their orientations. To motivate a precise definition of this type of topological similarity we first formalize the properties of a time reparametrization.

Let ϕ^t be a flow on a topological space X. We seek a function $\rho : \mathbb{R} \times X \to \mathbb{R}$ so that for each $x \in X$ the function $\rho(\cdot, x) : \mathbb{R} \to \mathbb{R}$ rescales or translates time along the orbit of x. First, for each $x \in X$ we would like $\phi^{\rho(\cdot,x)}(x)$ to map $\mathbb{R}$ onto the orbit of x. Consequently, we require that $\rho(\cdot, x)$ be surjective for each $x \in X$. Second, for each $x \in X$ we would like the image of $\phi^{\rho(\cdot,x)}(x)$ to have the same orientation as the orbit of x. So, we require that $\rho(\cdot, x)$ be strictly increasing for each $x \in X$.

Definition 1.3.1. Let X be a set. A function $\rho : \mathbb{R} \times X \to \mathbb{R}$ is called a **reparametrization** if

(i) $\rho(\cdot, x) : \mathbb{R} \to \mathbb{R}$ is surjective, and

(ii) $\rho(\cdot, x) : \mathbb{R} \to \mathbb{R}$ is strictly increasing

for each $x \in X$.

Example 1.3.2. Define $\rho : \mathbb{R} \times \mathbb{R} \to \mathbb{R}$ by

$$\rho(t, x) = \begin{cases} t & \text{if } x < 0, \\ t + 1 & \text{if } x \geq 0. \end{cases}$$

For each $x \in \mathbb{R}$, the function $\rho(\cdot, x) : \mathbb{R} \to \mathbb{R}$ is surjective and strictly increasing. Therefore, ρ is a reparametrization.

Example 1.3.3. Define $\rho : \mathbb{R} \times \mathbb{R} \to \mathbb{R}$ by

$$\rho(t, x) = \begin{cases} t & \text{if } x < 0, \\ 2t & \text{if } x \geq 0. \end{cases}$$

For each $x \in \mathbb{R}$, the function $\rho(\cdot, x) : \mathbb{R} \to \mathbb{R}$ is surjective and strictly increasing. Therefore, ρ is a reparametrization.

Example 1.3.4. Let X be a set, and let s be a positive real number. Define $\rho : \mathbb{R} \times X \to \mathbb{R}$ by

$$\rho(t, x) = st.$$

Since s is positive, and ρ is linear as a function of t, the function ρ is a reparametrization.

Although the next reparametrization may appear esoteric, we shall use it in Examples 1.3.9 and 2.5.13 to develop a very important example.

Example 1.3.5. Let $\mathbb{D}$ be the closed unit disk in the complex plane. Define $\rho : \mathbb{R} \times \text{int}(\mathbb{D}) \to \mathbb{R}$ by

$$\rho(t, z) = \int_0^t \sin^2(\tau + \arg z) + 1 - \left[\frac{|z|e^\tau}{1 + |z|(e^\tau - 1)}\right]^2 d\tau.$$

By the Fundamental Theorem of Calculus, $\rho(\cdot, z) : \mathbb{R} \to \mathbb{R}$ is differentiable. Since

$$\frac{\partial \rho}{\partial t}(t, z) = \sin^2(t + \arg z) + 1 - \left[\frac{|z|e^t}{1 + |z|(e^t - 1)}\right]^2 > 0,$$

the function $\rho(\cdot, z) : \mathbb{R} \to \mathbb{R}$ is strictly increasing.

Since

$$1 - |z|^2 \leq \sin^2(\tau + \arg z) + 1 - \left[\frac{|z|e^\tau}{1 + |z|(e^\tau - 1)}\right]^2$$

for all $z \in \operatorname{int}(\mathbb{D})$, the comparison test for improper integrals implies that

$$\int_0^t \sin^2(\tau + \arg z) + 1 - \left[\frac{|z|e^\tau}{1 + |z|(e^\tau - 1)}\right]^2 d\tau \to \pm\infty$$

as $t \to \pm\infty$, respectively. Thus, the continuity of $\rho(\cdot, z) : \mathbb{R} \to \mathbb{R}$ implies the surjectivity of $\rho(\cdot, z) : \mathbb{R} \to \mathbb{R}$. Therefore, ρ is a reparametrization.

Example 1.3.6. Let ϕ^t be a flow on a topological space X. Let $\beta : X \to \mathbb{R}$ be a continuous positive function such that

$$\int_0^t \beta(\phi^\tau(x))\, d\tau \to \pm\infty$$

as $t \to \pm\infty$, respectively, for each $x \in X$. Define $\rho : \mathbb{R} \times X \to \mathbb{R}$ by

$$\rho(t, x) = \int_0^t \beta(\phi^\tau(x))\, d\tau.$$

Let $x \in X$. By the Fundamental Theorem of Calculus,

$$\frac{\partial \rho}{\partial t}(t, x) = \beta(\phi^t(x)) > 0.$$

Thus, $\rho(\cdot, x) : \mathbb{R} \to \mathbb{R}$ is continuous and strictly increasing. Since

$$\int_0^t \beta(\phi^\tau(x))\, d\tau \to \pm\infty$$

as $t \to \pm\infty$, respectively, the function $\rho(\cdot, x) : \mathbb{R} \to \mathbb{R}$ is surjective. Therefore, ρ is a reparametrization.

Having developed the concept of reparametrization, we now define a second mode of topological similarity.

Definition 1.3.7. Let X and Y be topological spaces. Two flows $\phi^t : X \to X$ and $\psi^t : Y \to Y$ are **topologically equivalent** if there exists a homeomorphism $h : X \to Y$ and a reparametrization $\rho : \mathbb{R} \times X \to \mathbb{R}$ such that

$$\psi^t(h(x)) = h(\phi^{\rho(t,x)}(x))$$

for all $x \in X$ and $t \in \mathbb{R}$. The ordered pair (h, ρ) is a **topological equivalence** from ϕ^t to ψ^t.

A topological conjugacy $h : X \to Y$ corresponds to a topological equivalence (h, ρ) with $\rho(t, x) = t$ for all $t \in \mathbb{R}$ and $x \in X$. Thus, topologically conjugate flows are topologically equivalent.

Example 1.3.8. Let A be an $n \times n$ matrix with real entries. Consider the linear differential equation

$$\dot{x} = Ax$$

on $\mathbb{R}^n$. By Example 1.1.6 the flow $\phi^t : \mathbb{R}^n \to \mathbb{R}^n$ associated to this differential equation is

$$\phi^t(x) = e^{At}x.$$

Let s be a positive real number. Corresponding to the differential equation

$$\dot{x} = (sA)x$$

is the flow $\psi^t : \mathbb{R}^n \to \mathbb{R}^n$ defined by

$$\psi^t(x) = e^{(sA)t}x.$$

According to Example 1.3.4 the function $\rho : \mathbb{R} \times \mathbb{R}^n \to \mathbb{R}$ defined by

$$\rho(t, x) = st$$

is a reparametrization. Thus,

$$\psi^t(x) = e^{(sA)t}x = e^{A(st)}x = \phi^{\rho(t,x)}(x).$$

Therefore, if $h : \mathbb{R}^n \to \mathbb{R}^n$ is the identity map, then (h, ρ) is a topological equivalence from ϕ^t to ψ^t.

Example 1.3.9. Let $\mathbb{D}$ be the closed unit disk in the complex plane. Recall the relation $z = re^{i\theta}$ between a complex number z and its polar coordinates (r, θ).

Consider the flow ϕ^t corresponding to the system of differential equations

$$\begin{aligned} \dot{r} &= r(1-r), \\ \dot{\theta} &= 1 \end{aligned}$$

in polar coordinates on $\operatorname{int}(\mathbb{D})$. Notice that this system consists of the logistic differential equation in the radial coordinate r and a constant differential equation in the angular coordinate θ. We shall refer to ϕ^t as the **logistic rotation-dilation**.

Let ψ^t be the flow associated with the system of differential equations

$$\begin{aligned} \dot{r} &= r(1-r)(\sin^2\theta + 1 - r^2), \\ \dot{\theta} &= \sin^2\theta + 1 - r^2 \end{aligned}$$

in polar coordinates on $\operatorname{int}(\mathbb{D})$. We shall refer to ψ^t as the **arrested logistic rotation-dilation**.

According to Example 1.3.5 the function $\rho : \mathbb{R} \times \mathrm{int}(\mathbb{D}) \to \mathbb{R}$ defined by

$$\rho(t, z) = \int_0^t \sin^2(\tau + \arg z) + 1 - \left[\frac{|z|e^\tau}{1 + |z|(e^\tau - 1)}\right]^2 d\tau$$

is a reparametrization. In polar coordinates,

$$\rho(t, (r, \theta)) = \int_0^t \sin^2(\tau + \theta) + 1 - \left[\frac{re^\tau}{1 + r(e^\tau - 1)}\right]^2 d\tau.$$

By the Chain Rule,

$$\begin{aligned}
\left.\frac{d}{dt}\right|_{t=0} \phi^{\rho(t,(r,\theta))}(r, \theta) &= \left.\frac{d}{dt}\right|_{t=0} \phi(\rho(t, (r, \theta)), (r, \theta)) \\
&= \frac{d\phi}{d\rho}(\rho(0, (r, \theta)), (r, \theta)) \frac{d\rho}{dt}(0, (r, \theta)) \\
&= \left.\frac{d}{dt}\right|_{t=0} \phi^t(r, \theta) \frac{d\rho}{dt}(0, (r, \theta)) \\
&= \begin{pmatrix} r(1 - r) \\ 1 \end{pmatrix} (\sin^2 \theta + 1 - r^2) \\
&= \begin{pmatrix} r(1 - r)(\sin^2 \theta + 1 - r^2) \\ \sin^2 \theta + 1 - r^2 \end{pmatrix} \\
&= \left.\frac{d}{dt}\right|_{t=0} \psi^t(r, \theta)
\end{aligned}$$

for each point in $\mathrm{int}(\mathbb{D})$ with polar coordinates (r, θ). Since $\phi^0(r, \theta) = (r, \theta) = \psi^0(r, \theta)$,

$$\psi^t(r, \theta) = \phi^{\rho(t,(r,\theta))}(r, \theta)$$

for all real numbers t and all points in $\mathrm{int}(\mathbb{D})$ with polar coordinates (r, θ). Therefore, if $h : \mathrm{int}(\mathbb{D}) \to \mathrm{int}(\mathbb{D})$ is the identity map, then (h, ρ) is a topological equivalence from ϕ^t to ψ^t.

Example 1.3.10. Let ϕ^t be the flow of a smooth vector field f on a manifold M. Let $\beta : M \to \mathbb{R}$ be a smooth positive function such that

$$\int_0^t \beta(\phi^\tau(x))\, d\tau \to \pm\infty$$

as $t \to \pm\infty$, respectively, for each $x \in M$. We will show that scaling the vector field f by the real-valued function β generates a flow which is topologically equivalent to ϕ^t.

Define $\rho : \mathbb{R} \times M \to \mathbb{R}$ by

$$\rho(t, x) = \int_0^t \beta(\phi^\tau(x))\, d\tau.$$

By Example 1.3.6 the function ρ is a reparametrization. The Fundamental Theorem of Calculus guarantees that $\rho(\cdot, x) : \mathbb{R} \to \mathbb{R}$ is smooth for each $x \in M$. By the Chain Rule,

$$\left.\frac{d}{dt}\right|_{t=0} \phi^{\rho(t,x)}(x) = \left.\frac{d}{dt}\right|_{t=0} \phi^t(x) \left.\frac{d}{dt}\right|_{t=0} \rho(t,x) = f(x)\beta(x) = (\beta f)(x)$$

for all $x \in M$.

Let ψ^t be the flow of the vector field βf on M. Then

$$\left.\frac{d}{dt}\right|_{t=0} \psi^t(x) = (\beta f)(x)$$

for all $x \in M$. Thus,

$$\left.\frac{d}{dt}\right|_{t=0} \psi^t(x) = \left.\frac{d}{dt}\right|_{t=0} \phi^{\rho(t,x)}(x)$$

for all $x \in M$. Since $\psi^0(x) = \phi^{\rho(0,x)}(x)$ for all $x \in M$,

$$\psi^t(x) = \phi^{\rho(t,x)}(x)$$

for all $t \in \mathbb{R}$ and all $x \in M$. Therefore, ϕ^t is topologically equivalent to ψ^t.

If $\rho : \mathbb{R} \times X \to \mathbb{R}$ is a reparametrization, then Definition 1.3.1 implies that $\rho(\cdot, x) : \mathbb{R} \to \mathbb{R}$ is continuous for each $x \in X$. However, Definition 1.3.1 does not require that $\rho : \mathbb{R} \times X \to \mathbb{R}$ be continuous or respect the group property of any particular flow $\phi^t : X \to X$. Consequently, as the following examples demonstrate, the function $\Phi : \mathbb{R} \times X \to X$ defined by

$$\Phi(t,x) = \phi^{\rho(t,x)}(x)$$

is not necessarily a flow on X.

Example 1.3.11. Consider the flow $\phi^t : \mathbb{R} \to \mathbb{R}$ defined by

$$\phi^t(x) = x + t.$$

This is the flow of Example 1.1.5 with $n = 1$ and $c = 1$.

By Example 1.3.2 the function $\rho : \mathbb{R} \times \mathbb{R} \to \mathbb{R}$ defined by

$$\rho(t,x) = \begin{cases} t & \text{if } x < 0, \\ t+1 & \text{if } x \geq 0. \end{cases}$$

is a reparametrization. Define $\Phi : \mathbb{R} \times \mathbb{R} \to \mathbb{R}$ by

$$\Phi(t,x) = \phi^{\rho(t,x)}(x).$$

Since

$$\begin{aligned} \phi^{\rho(t,x)}(x) &= \begin{cases} \phi^t(x) & \text{if } x < 0, \\ \phi^{t+1}(x) & \text{if } x \geq 0, \end{cases} \\ &= \begin{cases} x + t & \text{if } x < 0, \\ x + t + 1 & \text{if } x \geq 0 \end{cases} \end{aligned}$$

is not continuous, the function Φ is not a flow.

Even a continuous reparametrization may not transform a flow into a flow.

Example 1.3.12. Consider the flow $\phi^t : \mathbb{R} \to \mathbb{R}$ defined by

$$\phi^t(x) = x + t.$$

Again, this is the flow of Example 1.1.5 with $n = 1$ and $c = 1$.

By Example 1.3.3 the function $\rho : \mathbb{R} \times \mathbb{R} \to \mathbb{R}$ defined by

$$\rho(t, x) = \begin{cases} t & \text{if } x < 0, \\ 2t & \text{if } x \geq 0 \end{cases}$$

is a reparametrization. Define $\Phi : \mathbb{R} \times \mathbb{R} \to \mathbb{R}$ by

$$\Phi(t, x) = \phi^{\rho(t,x)}(x).$$

Since

$$\Phi(1, 0) = \phi^{\rho(1,0)}(0) = \phi^2(0) = 2,$$

we obtain

$$\Phi(1, \Phi(1, -1)) = \Phi(1, \phi^{\rho(1,-1)}(-1)) = \Phi(1, \phi^1(-1)) = \Phi(1, 0) = 2.$$

On the other hand,

$$\Phi(1 + 1, -1) = \phi^{\rho(2,-1)}(-1) = \phi^2(-1) = 1.$$

Since

$$\Phi(1 + 1, -1) \neq \Phi(1, \Phi(1, -1)),$$

the function Φ does not satisfy the group property of flows. Therefore, Φ is not a flow.

If a reparametrization does indeed transform a flow ϕ^t into a flow Φ^t, then we call Φ^t a time change of ϕ^t.

Definition 1.3.13. Let ϕ^t be a flow on a topological space X. A function $\Phi : \mathbb{R} \times X \to X$ is a **time change** of ϕ^t if there exists a reparametrization $\rho : \mathbb{R} \times X \to \mathbb{R}$ such that $\Phi(t, x) = \phi^{\rho(t,x)}(x)$ and Φ is a flow.

The next proposition relates topological equivalences to time changes.

Proposition 1.3.14. *Let X and Y be topological spaces. If (h, ρ) is a topological equivalence from a flow $\phi^t : X \to X$ to a flow $\psi^t : Y \to Y$, then the function $\Phi : \mathbb{R} \times X \to X$ defined by $\Phi(t, x) = \phi^{\rho(t,x)}(x)$ is a time change of ϕ^t.*

Proof. Applying the definition of Φ and the topological equivalence (h, ρ) from ϕ^t to ψ^t, we have

$$h(\Phi(t,x)) = \psi^t(h(x))$$

for all $t \in \mathbb{R}$ and all $x \in X$.

First, since h is a homeomorphism and ψ is continuous,

$$\Phi(t,x) = h^{-1}(\psi(t, h(x)))$$

is continuous.

Second, since h is a homeomorphism and ψ^t is a homeomorphism for each $t \in \mathbb{R}$,

$$\Phi(t,x) = h^{-1}(\psi^t(h(x)))$$

is a homeomorphism for each $t \in \mathbb{R}$.

Finally, we demonstrate the group property. Let $s, t \in \mathbb{R}$. By the definition of topological equivalence and the group property applied to ψ^t,

$$\begin{aligned}
\Phi(t+s,x) &= h^{-1}(\psi^{t+s}(h(x))) \\
&= h^{-1}(\psi^s(\psi^t(h(x)))) \\
&= h^{-1}(h(\Phi(s, h^{-1}(\psi^t(h(x)))))) \\
&= \Phi(s, (h^{-1}(h(\Phi(t, h^{-1}(h(x))))))) \\
&= \Phi(s, \Phi(t,x)).
\end{aligned}$$

Thus, Φ is a flow. Therefore, Φ^t is a time change of ϕ^t. □

From Proposition 1.3.14 we obtain an important characterization of topologically equivalent flows: the flows ϕ^t and ψ^t are topologically equivalent if and only if there is a time change of ϕ^t which is topologically conjugate to ψ^t.

We now demonstrate that if X and Y are topological spaces, and (h, ρ) is a topological equivalence from a flow $\phi^t : X \to X$ to a flow $\psi^t : Y \to Y$, then there is a topological equivalence from ψ^t to ϕ^t.

Definition 1.3.1 implies that $\rho(\cdot, x) : \mathbb{R} \to \mathbb{R}$ is invertible for each $x \in X$. Therefore, for each $s \in \mathbb{R}$ and $y \in Y$ there exists a unique $t \in \mathbb{R}$ such that $s = \rho(t, h^{-1}(y))$. Define $\tilde{\rho} : \mathbb{R} \times Y \to \mathbb{R}$ by

$$\tilde{\rho}(s,y) = t.$$

The following proposition shows that $(h^{-1}, \tilde{\rho})$ is a topological equivalence from ψ^t to ϕ^t.

Proposition 1.3.15. *Let X and Y be topological spaces. If (h, ρ) is a topological equivalence from a flow $\phi^t : X \to X$ to a flow $\psi^t : Y \to Y$, then the function $\tilde{\rho} : \mathbb{R} \times Y \to \mathbb{R}$ defined by*

$$\tilde{\rho}(s,y) = t$$

where $s = \rho(t, h^{-1}(y))$ is a reparametrization, and $(h^{-1}, \tilde{\rho})$ is a topological equivalence from ψ^t to ϕ^t.

Proof. To show that $\tilde{\rho}$ is a reparametrization of ψ^t we verify properties (i) and (ii) of Definition 1.3.1. Let $y \in Y$.

(i) Let $t \in \mathbb{R}$. By the definition of $\tilde{\rho}$,

$$\tilde{\rho}(\rho(t, h^{-1}(y)), y) = t.$$

Therefore, $\tilde{\rho}(\cdot, y) : \mathbb{R} \to \mathbb{R}$ is surjective.

(ii) Let $s, t \in \mathbb{R}$ such that $s > t$. If $u = \tilde{\rho}(s, y)$, then $s = \rho(u, h^{-1}(y))$, and if $v = \tilde{\rho}(t, y)$, then $t = \rho(v, h^{-1}(y))$. Since $s > t$, we obtain

$$\rho(u, h^{-1}(y)) > \rho(v, h^{-1}(y)).$$

Because $\rho(\cdot, h^{-1}(y)) : \mathbb{R} \to \mathbb{R}$ is strictly increasing, $u > v$. Thus,

$$\tilde{\rho}(s, y) > \tilde{\rho}(t, y).$$

Therefore, $\tilde{\rho}(\cdot, y) : \mathbb{R} \to \mathbb{R}$ is strictly increasing. Consequently, $\tilde{\rho}$ is a reparametrization of ψ^t.

Now we show that the pair $(h^{-1}, \tilde{\rho})$ is a topological equivalence from ψ^t to ϕ^t. Let $y \in Y$. There exists $x \in X$ such that $y = h(x)$. Let $s \in \mathbb{R}$, and define $t = \tilde{\rho}(s, y)$. Since (h, ρ) is a topological equivalence from ϕ^t to ψ^t,

$$\psi^t(h(x)) = h(\phi^{\rho(t,x)}(x)).$$

Applying h^{-1} to each side we obtain

$$h^{-1}(\psi^t(y)) = \phi^{\rho(t,x)}(h^{-1}(y)).$$

Thus,

$$h^{-1}(\psi^{\tilde{\rho}(s,y)}(y)) = \phi^s(h^{-1}(y)).$$

Therefore, $(h^{-1}, \tilde{\rho})$ is a topological equivalence from ψ^t to ϕ^t. □

An ambitious program has been to classify all flows on a given phase space up to topological equivalence. Kneser [**27**], Kaplan [**24**], Markus [**29**] and Whitney [**45**] contain results for surface flows.

1.4. Invariant Sets

Some important sets related to a flow ϕ^t on a space X are subsets A of X with the property that if $x \in A$, then $\mathcal{O}(x) \subseteq A$. We call these sets invariant with respect to the flow.

Definition 1.4.1. Let ϕ^t be a flow on a topological space X. A subset A of X is **positively invariant** with respect to ϕ^t if $\phi^t(A) \subseteq A$ for all positive real numbers t. A subset A of X is **negatively invariant** with respect to ϕ^t if $\phi^{-t}(A) \subseteq A$ for all positive real numbers t. A subset A of X is **invariant** with respect to ϕ^t if A is positively invariant with respect to ϕ^t and A is negatively invariant with respect to ϕ^t.

Clearly, a set A is invariant with respect to a flow ϕ^t if $\phi^t(A) \subseteq A$ for all real numbers t. If A is invariant with respect to ϕ^t, then ϕ^t restricted to A is a flow on A.

Example 1.4.2. If ϕ^t is a flow on a topological space X, then $\emptyset$ and X are invariant sets with respect to ϕ^t.

Example 1.4.3. A forward orbit is a positively invariant set. A backward orbit is a negatively invariant set. An orbit is an invariant set.

Complements, intersections and unions of invariant sets are invariant.

Proposition 1.4.4.
(i) *The relative complement of any set which is (positively, negatively) invariant with respect to a flow is (negatively, positively) invariant with respect to the flow.*
(ii) *The intersection of any collection of sets which are invariant with respect to a flow is invariant with respect to the flow.*
(iii) *The union of any collection of sets which are invariant with respect to a flow is invariant with respect to the flow.*

Proof. Let ϕ^t be a flow on a topological space X.

(i) Let A be a subset of X which is positively invariant with respect to ϕ^t. We proceed by proving the contrapositive. If $X \setminus A$ is not negatively invariant with respect to ϕ^t, then there exists $x \in X \setminus A$ and $\tau < 0$ such that $\phi^\tau(x) \notin X \setminus A$. That is, $\phi^\tau(x) \in A$. Consequently, $x = \phi^{-\tau}(\phi^\tau(x)) \in X \setminus A$. Thus, A is not positively invariant. Therefore, if A is positively invariant with respect to ϕ^t, then $X \setminus A$ is negatively invariant with respect to ϕ^t. The proofs of the other two statements are similar.

(ii) Let $\{A_\lambda\}_{\lambda \in \Lambda}$ be a collection of subsets of X which are invariant with respect to ϕ^t. Let $x \in \bigcap_{\lambda \in \Lambda} A_\lambda$. Let $\tau \in \mathbb{R}$. For each $\lambda \in \Lambda$ the point $\phi^\tau(x) \in A_\lambda$ since $x \in A_\lambda$ and A_λ is invariant. Thus, $\phi^\tau(x) \in \bigcap_{\lambda \in \Lambda} A_\lambda$. Therefore, $\bigcap_{\lambda \in \Lambda} A_\lambda$ is invariant with respect to ϕ^t.

(iii) Apply De Morgan's Law and parts (i) and (ii). □

The next proposition guarantees that the closure of an invariant set is invariant.

Proposition 1.4.5. (i) *The closure of a set which is positively invariant with respect to a flow is positively invariant with respect to the flow.*

(ii) *The closure of a set which is negatively invariant with respect to a flow is negatively invariant with respect to the flow.*

(iii) *The closure of a set which is invariant with respect to a flow is invariant with respect to the flow.*

Proof. Let ϕ^t be a flow on a topological space X.

(i) If A is positively invariant with respect to ϕ^t, then it suffices to show that the set of limit points of A is positively invariant with respect to ϕ^t. Let x be a limit point of A, let $\tau > 0$, and let U be an open set in X containing $\phi^\tau(x)$. By the continuity of ϕ^τ, the set $\phi^{-\tau}(U)$ is open and contains x. Since x is a limit point of A, there exists $y \in A \cap \phi^{-\tau}(U)$. Because A is positively invariant, $\phi^\tau(y) \in A \cap U$. Thus, $\phi^\tau(x)$ is a limit point of A. Therefore, $\mathrm{cl}(A)$ is positively invariant with respect to ϕ^t.

(ii) The proof is similar to (i).

(iii) If A is invariant with respect to ϕ^t, then A is positively invariant with respect to ϕ^t and negatively invariant with respect to ϕ^t. By (i) and (ii), $\mathrm{cl}(A)$ is positively invariant with respect to ϕ^t and negatively invariant with respect to ϕ^t. Therefore, $\mathrm{cl}(A)$ is invariant with respect to ϕ^t. □

Topological equivalences carry invariant sets to invariant sets.

Proposition 1.4.6. *Let X and Y be topological spaces. Let (h, ρ) be a topological equivalence from a flow $\phi^t : X \to X$ to a flow $\psi^t : Y \to Y$. If A is invariant with respect to ϕ^t, then $h(A)$ is invariant with respect to ψ^t.*

Proof. If $y \in h(A)$, then there exists $x \in A$ such that $y = h(x)$. Let $\tau \in \mathbb{R}$. Applying the topological equivalence (h, ρ) and the invariance of A with respect to ϕ^t,

$$\psi^\tau(y) = \psi^\tau(h(x)) = h(\phi^{\rho(\tau,x)}(x)) \in h(A).$$

Therefore, $h(A)$ is invariant with respect to ψ^t. □

1.5. Exercises

(1) Consider the flow ϕ^t corresponding to the differential equation

$$\dot{x} = x$$

on $\mathbb{R}$.
   (a) Find an explicit expression for ϕ^t.
   (b) Find the orbits of ϕ^t.
   (c) Find an explicit expression for the time-one map of ϕ^t.
(2) Consider the flow ϕ^t corresponding to the differential equation

$$\dot{x} = -x$$

on $\mathbb{R}$.
   (a) Find an explicit expression for ϕ^t.
   (b) Find the orbits of ϕ^t.
   (c) Find an explicit expression for the time-one map of ϕ^t.
(3) Consider the flow ϕ^t corresponding to the system of differential equations

$$\begin{aligned} \dot{x} &= -y, \\ \dot{y} &= x \end{aligned}$$

on $\mathbb{R}^2$.
   (a) Find an explicit expression for ϕ^t.
   (b) Find the orbits of ϕ^t.
   (c) Find an explicit expression for the time-one map of ϕ^t.
(4) Prove that the following differential equations do not define flows on $\mathbb{R}$.
   (a) $\dot{x} = x^{\frac{2}{3}}$.
   (b) $\dot{x} = x^2$.
   (c) $\dot{x} = 1 + x^2$.
(5) An **action** of a group G on a set S is a function $G \times S \to S$, denoted $(g, x) \mapsto gx$, such that for every $x \in S$ and $g_1, g_2 \in G$ we have $ex = x$ and $(g_1 g_2)x = g_1(g_2 x)$ where e is the identity element of G. The **orbit** of $x \in S$ is the set $\text{orb}(x) = \{gx \mid g \in G\}$. Let ϕ^t be a flow on a topological space X.
   (a) Prove that ϕ is a continuous action of the topological group $(\mathbb{R}, +)$ on X.
   (b) Prove that $\text{orb}(x) = \mathcal{O}(x)$ for all $x \in X$.
(6) Let X be a topological space. A continuous function $f : X \to X$ is **homotopic to the identity** if there exists a continuous function $H : X \times [0, 1] \to X$ such that $H(x, 0) = f(x)$ and $H(x, 1) = x$ for all $x \in X$. Prove that the time-one map of a flow is homotopic to the identity.
(7) Let X and Y be topological spaces. Let $h : X \to Y$ be a topological conjugacy from a flow $\phi^t : X \to X$ to a flow $\psi^t : Y \to Y$. Prove that if $x \in X$, then $h(\mathcal{O}(x, \phi^t)) = \mathcal{O}(h(x), \psi^t)$.

(8) Let (h, ρ) be a topological equivalence from a flow $\phi^t : X \to X$ to a flow $\psi^t : Y \to Y$. Prove that if $A \subseteq X$ is invariant with respect to ϕ^t, then $(h|_A, \rho|_{\mathbb{R}\times A})$ is a topological equivalence from $\phi^t|_A$ to $\psi^t|_{h(A)}$.

(9) Let X and Y be topological spaces. Prove that if $\phi^t : X \to X$ and $\psi^t : Y \to Y$ are topologically equivalent flows, then there exists a topological equivalence (h, ρ) from ϕ^t to ψ^t such that $\rho(0, x) = 0$ for all $x \in X$.

(10) Prove Proposition 1.4.4 (iii) without using Proposition 1.4.4 (ii).

(11) Prove Proposition 1.4.5 (ii).

(12) Let X and Y be topological spaces. Prove that if (h, ρ) is a topological equivalence from a flow $\phi^t : X \to X$ to a flow $\psi^t : Y \to Y$ and A is a positively invariant set with respect to ϕ^t, then $h(A)$ is positively invariant with respect to ψ^t.

Chapter 2

Recurrent Points

A major objective of the study of dynamical systems is to describe the eventual behavior of the orbits of a map or flow. If $\phi^t : X \to X$ is a flow, then our goal is to understand the disposition of $\phi^t(x)$ as $t \to \pm\infty$ for each $x \in X$.

Solutions of differential equations often tend toward equilibrium solutions or periodic solutions in forward or backward time. An equilibrium solution or periodic solution exhibits recurrence in the sense that it returns to its initial condition infinitely often. However, solutions of differential equations may also tend toward solutions displaying more complicated recurrent behavior. To describe these more complicated types of asymptotic behavior we seek to understand more intricate forms of recurrence.

This chapter develops several topological modes of recurrence. Ideally, we seek a definition of recurrence so that the set $R(\phi^t)$ of recurrent points with respect to a flow $\phi^t : X \to X$ has six desirable properties:

(1) **Flow Invariance Property.** *The set* $R(\phi^t)$ *is an invariant set with respect to* ϕ^t. Recurrence should be a property of an orbit rather than a point.

(2) **Closure Property.** *The set* $R(\phi^t)$ *is closed with respect to the topology of* X. Limit points of recurrent points should be recurrent.

(3) **Topological Invariance Property.** *If* $h : X \to Y$ *is a topological conjugacy from* ϕ^t *to a flow* $\psi^t : Y \to Y$, *then*

$$R(\psi^t) = h(R(\phi^t)).$$

Topologically conjugate flows should have recurrent sets which map to one another under a conjugacy. That is, recurrence should be

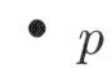

Figure 2.1. A fixed point

topological for our purposes. We also examine the invariance of sets of recurrent points with respect to topological equivalences.

(4) **Accumulation Property.** *If $x \in X$ and there exists a compact subset of X containing $\mathcal{O}(x)$, then*

$$\lim_{n\to\infty} \phi^{t_n}(x) \in R(\phi^t)$$

for every sequence t_n of real numbers such that $t_n \to \pm\infty$ as $n \to \infty$ whenever this limit exists. The set $R(\phi^t)$ should reflect the eventual behavior of every bounded orbit of ϕ^t, so that the asymptotic analysis of ϕ^t reduces to the analysis of ϕ^t restricted to $R(\phi^t)$.

(5) **Restriction Property.** $R(\phi^t|_{R(\phi^t)}) = R(\phi^t)$. The property of being recurrent should be intrinsic to the set of recurrent points.

(6) **Decomposition Property.** *There exists a partition of $R(\phi^t)$ into closed invariant irreducible subsets.* The set $R(\phi^t)$ should separate into irreducible components where the meaning of irreducibility depends on the type of recurrence defined by $R(\phi^t)$. Combined with the Accumulation Property the asymptotic analysis of ϕ^t should reduce to the analysis of ϕ^t restricted to $R(\phi^t)$ which, in turn, should reduce to the analysis of ϕ^t on each irreducible subset of $R(\phi^t)$. Chapter 3 develops topological modes of irreducibility.

Let us emphasize that $R(\phi^t)$ is an arbitrary set whose properties we shall explore for several possible definitions of $R(\phi^t)$. For example, in Section 2.1 we shall consider the case in which $R(\phi^t)$ is the set of fixed points of ϕ^t, and in Section 2.2 we shall consider the case in which $R(\phi^t)$ is the set of periodic points of ϕ^t.

Having discussed the desirable properties of sets of recurrent points, we proceed to systematically develop six types of recurrence, comparing the properties of each type with these desirable properties.

2.1. Fixed Points

A point fixed by a flow for all time exhibits the simplest type of recurrence.

Definition 2.1.1. Let ϕ^t be a flow on a topological space X. A point $p \in X$ is a **fixed point** of ϕ^t if $\phi^t(p) = p$ for all $t \in \mathbb{R}$. The **fixed set** of ϕ^t, denoted $\mathrm{Fix}(\phi^t)$, is the set of all fixed points of ϕ^t. See Figure 2.1.

The orbit of a fixed point p is $\{p\}$. By Proposition 1.1.12, if the orbit of a point x contains a fixed point p, then $\mathcal{O}(x) = \{p\}$. In particular, $x = p$.

The fixed set of a flow satisfies the Flow Invariance Property. Despite the triviality of its proof, we formally state and prove this fact in the interest of completeness.

Proposition 2.1.2. *The fixed set of a flow is invariant with respect to the flow.*

Proof. Let ϕ^t be a flow. If $p \in \text{Fix}(\phi^t)$, then $\phi^\tau(p) = p \in \text{Fix}(\phi^t)$ for each real number τ. Therefore, $\text{Fix}(\phi^t)$ is invariant with respect to ϕ^t. $\square$

To facilitate providing examples of fixed points, we establish a connection between the zeros of a smooth vector field and the fixed points of its corresponding flow.

Proposition 2.1.3. *Let f be a smooth vector field on manifold M. If $\phi^t : M \to M$ is a smooth flow satisfying*

$$\left.\frac{d}{dt}\right|_{t=0} \phi^t(x) = f(x)$$

for all $x \in M$, then $\text{Fix}(\phi^t) = \{p \in M \mid f(p) = 0\}$.

Proof. If p is a fixed point of ϕ^t, then $\phi^t(p) = p$ for all $t \in \mathbb{R}$, and

$$f(p) = \left.\frac{d}{dt}\right|_{t=0} \phi^t(p) = 0.$$

Conversely, if $p \in M$ and $f(p) = 0$, then $\phi^t(p) = p$ for all $t \in \mathbb{R}$ is a solution to the differential equation $\dot{x} = f(x)$ with the initial condition $x(0) = p$. Since f is smooth, the existence and uniqueness theorem for ordinary differential equations guarantees that this solution is unique. Thus, p is a fixed point of ϕ^t. Therefore, the fixed points of ϕ^t are exactly the zeros of f. $\square$

Example 2.1.4. Let A be an $n \times n$ matrix with real entries. By Proposition 2.1.3 the fixed set of the flow corresponding to the linear differential equation

$$\dot{x} = Ax$$

on $\mathbb{R}^n$ is the kernel of the linear transformation which A represents.

Example 2.1.5. Consider the logistic differential equation

$$\dot{x} = x(1 - x)$$

on $[0, 1]$. By Proposition 2.1.3 the fixed points of the corresponding flow are exactly the zeros of the function $f(x) = x(1 - x)$. Therefore,

$$\text{Fix}(\phi^t) = \{0, 1\}.$$

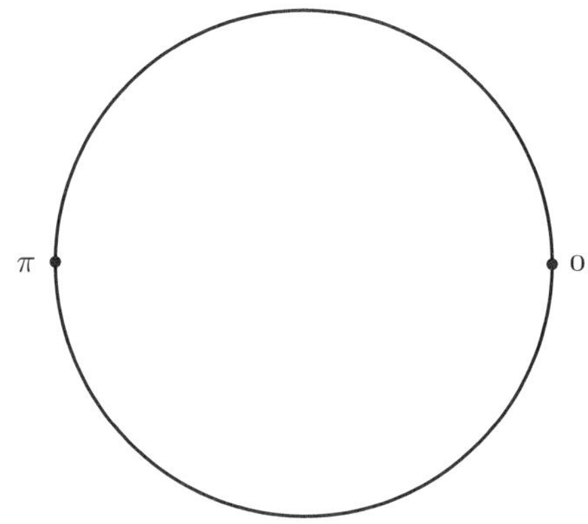

Figure 2.2. The fixed points of the arrested rotation

Example 2.1.6. Consider the differential equation

$$\dot{\theta} = \sin^2 \theta$$

in the angular coordinate θ on S^1. The flow corresponding to this differential equation is the arrested rotation of Example 1.1.8. By Proposition 2.1.3 the fixed points of the arrested rotation are exactly the zeros of $f(\theta) = \sin^2 \theta$. Therefore, the fixed points of the flow are the points with angular coordinates 0 and π. See Figure 2.2.

In Proposition 2.1.3 the hypothesis that f is smooth implies that f is continuous. Consequently, $\mathrm{Fix}(\phi^t) = f^{-1}(\{0\})$ is a closed set. In general, the set of fixed points of a flow on a Hausdorff space satisfies the Closure Property.

Proposition 2.1.7. *The fixed set of a flow on a Hausdorff topological space is closed.*

Proof. Let ϕ^t be a flow on a Hausdorff topological space X. We shall prove that $X \setminus \mathrm{Fix}(\phi^t)$ is open. Let $x \in X \setminus \mathrm{Fix}(\phi^t)$. Thus, there exists a real number τ such that $\phi^\tau(x) \neq x$. Since X is Hausdorff, there exist disjoint open subsets U and V of X such that $\phi^\tau(x) \in U$ and $x \in V$.

Let $A = \phi^{-\tau}(U) \cap V$. Since ϕ^τ is continuous, A is an open set. We shall show that $A \subseteq X \setminus \mathrm{Fix}(\phi^t)$. If $y \in A$, then $\phi^\tau(y) \in U$ and $y \in V$. Because U and V are disjoint, $\phi^\tau(y) \neq y$. Hence, $y \in X \setminus \mathrm{Fix}(\phi^t)$. Consequently, A is an open set containing x, and $A \subseteq X \setminus \mathrm{Fix}(\phi^t)$. Thus, $X \setminus \mathrm{Fix}(\phi^t)$ is open. Therefore, $\mathrm{Fix}(\phi^t)$ is closed. □

Topological equivalences carry fixed sets to fixed sets. In particular, the fixed set of a flow satisfies the Topological Invariance Property.

Proposition 2.1.8. *Let X and Y be topological spaces. If (h, ρ) is a topological equivalence from the flow $\phi^t : X \to X$ to the flow $\psi^t : Y \to Y$, then*

$$Fix(\psi^t) = h(Fix(\phi^t)).$$

Proof. Let $x \in \mathrm{Fix}(\phi^t)$, and let $t \in \mathbb{R}$. Since (h, ρ) is a topological equivalence,

$$\psi^t(h(x)) = h(\phi^{\rho(t,x)}(x)) = h(x).$$

Consequently, $h(x) \in \mathrm{Fix}(\psi^t)$. Thus, $h(\mathrm{Fix}(\phi^t)) \subseteq \mathrm{Fix}(\psi^t)$.

To prove the reverse inclusion, Proposition 1.3.15 guarantees that there is a topological equivalence $(h^{-1}, \tilde{\rho})$ from ψ^t to ϕ^t. By the argument in the previous paragraph, $h^{-1}(\mathrm{Fix}(\psi^t)) \subseteq \mathrm{Fix}(\phi^t)$. Applying the bijection h to both sides of this inclusion we obtain $\mathrm{Fix}(\psi^t) \subseteq h(\mathrm{Fix}(\phi^t))$. Therefore,

$$\mathrm{Fix}(\psi^t) = h(\mathrm{Fix}(\phi^t)).$$ □

Example 2.3.8 will show that the fixed set does not satisfy the Accumulation Property. However, verifying that the fixed set satisfies the Restriction Property and the Decomposition Property is straightforward. We include statements and proofs in the interest of completeness.

Proposition 2.1.9. *If ϕ^t is a flow, then $Fix(\phi^t|_{Fix(\phi^t)}) = Fix(\phi^t)$.*

Proof. If $p \in \mathrm{Fix}(\phi^t|_{\mathrm{Fix}(\phi^t)})$, then $p \in \mathrm{Fix}(\phi^t)$. If $p \in \mathrm{Fix}(\phi^t)$, then

$$\phi^t|_{\mathrm{Fix}(\phi^t)}(p) = \phi^t(p) = p$$

for all $t \in \mathbb{R}$. So, $p \in \mathrm{Fix}(\phi^t|_{\mathrm{Fix}(\phi^t)})$. Therefore, $\mathrm{Fix}(\phi^t|_{\mathrm{Fix}(\phi^t)}) = \mathrm{Fix}(\phi^t)$. □

Proposition 2.1.10. *If ϕ^t is a flow and $\{p\}$ is closed for each $p \in Fix(\phi^t)$, then $Fix(\phi^t)$ is a union of disjoint closed sets which are invariant with respect to ϕ^t.*

Proof. By hypothesis $\{p\}$ is closed for each $p \in \mathrm{Fix}(\phi^t)$. The set $\mathcal{O}(p) = \{p\}$ is invariant with respect to ϕ^t for each $p \in \mathrm{Fix}(\phi^t)$. Therefore,

$$\mathrm{Fix}(\phi^t) = \bigcup_{p \in \mathrm{Fix}(\phi^t)} \{p\}$$

expresses $\mathrm{Fix}(\phi^t)$ as a union of disjoint closed invariant sets. □

Notice that the hypothesis of Proposition 2.1.10 holds for any flow on a Hausdorff space.

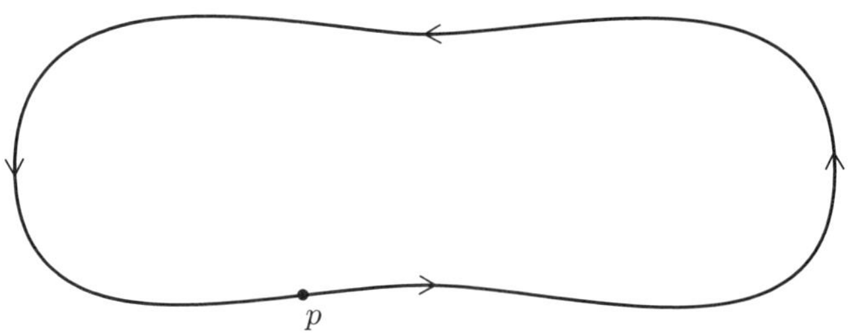

Figure 2.3. A periodic orbit

2.2. Periodic Points

A point carried back to itself by a flow exhibits the next most elementary type of recurrence.

Definition 2.2.1. Let ϕ^t be a flow on a topological space X. A point $p \in X$ is a **periodic point** of ϕ^t if there exists a positive real number τ such that $\phi^\tau(p) = p$. The **periodic set** of ϕ^t, denoted $\text{Per}(\phi^t)$, is the set of periodic points of ϕ^t. A **periodic orbit** of ϕ^t is an orbit of a periodic point of ϕ^t. See Figure 2.3.

If p is a periodic point of a flow ϕ^t, then the set $\{\tau > 0 \mid \phi^\tau(p) = p\}$ is nonempty and bounded below by 0. By the completeness of the real numbers the infimum of $\{\tau > 0 \mid \phi^\tau(p) = p\}$ exists.

Definition 2.2.2. Let p be a periodic point of a flow ϕ^t. The **period** of p is the real number $\inf\{\tau > 0 \mid \phi^\tau(p) = p\}$.

Example 2.2.3. If p is a fixed point of a flow ϕ^t, then $\phi^t(p) = p$ for all $t \in \mathbb{R}$. Thus, every fixed point is a periodic point with period 0. Therefore,

$$\text{Fix}(\phi^t) \subseteq \text{Per}(\phi^t).$$

Now we establish under very general hypotheses that if p is a periodic point of a flow ϕ^t with period T, then $\phi^T(p) = p$.

Proposition 2.2.4. *Let ϕ^t be a flow. If p is a periodic point of ϕ^t with period T and $\{p\}$ is a closed set, then $\phi^T(p) = p$.*

Proof. Let $A = \{\tau > 0 \mid \phi^\tau(p) = p\}$. By the definition of period, $T = \inf A$. So, $T \in \text{cl}(A)$. Because $\{p\}$ is closed, $(T, p) \in \text{cl}(A) \times \{p\} = \text{cl}(A \times \{p\})$.

Since ϕ is continuous and $\{p\}$ is closed,

$$\phi^T(p) = \phi(T, p) \in \mathrm{cl}(\phi(A \times \{p\})) = \mathrm{cl}(\{p\}) = \{p\}.$$

Therefore, $\phi^T(p) = p$. □

Proposition 2.2.5. *Let ϕ^t be a flow on a topological space X. If $p \in X$ and $\phi^\tau(p) = p$ for some real number τ, then $\phi^{n\tau}(p) = p$ for every integer n.*

Proof. If $n = 0$, then $\phi^0(p) = p$ by Proposition 1.1.2.

If n is a positive integer, then proceed by induction. If $n = 1$, then $\phi^\tau(p) = p$ by hypothesis. If $n \geq 2$, then assume as an inductive hypothesis that

$$\phi^{(n-1)\tau}(p) = p.$$

According to the group property of flows,

$$\phi^{n\tau}(p) = \phi^\tau(\phi^{(n-1)\tau}(p)) = \phi^\tau(p) = p.$$

By the Principle of Mathematical Induction, $\phi^{n\tau}(p) = p$ for every positive integer n.

If n is a negative integer, then $\phi^{-n\tau}(p) = p$ by the preceding argument. Applying $\phi^{n\tau}$ to both sides of this equation we obtain

$$p = \phi^{n\tau}(p)$$

by Proposition 1.1.2.

Therefore,

$$\phi^{n\tau}(p) = p$$

for every integer n. □

Corollary 2.2.6. *Let ϕ^t be a flow. If p is a periodic point of ϕ^t with period T and $\{p\}$ is a closed set, then $\phi^{nT}(p) = p$ for every integer n.*

Proof. Apply Propositions 2.2.4 and 2.2.5. □

Example 2.2.3 showed that every fixed point is a periodic point with period zero. The following proposition uses Proposition 2.2.5 to demonstrate the converse; every periodic point with period zero is a fixed point.

Proposition 2.2.7. *Let ϕ^t be a flow such that $\{\phi^t(p)\}$ is closed for each periodic point p of ϕ^t with period zero and each positive real number t. The set of fixed points of ϕ^t is the set of periodic points of ϕ^t with period zero.*

Proof. If p is a periodic point of ϕ^t with period 0, then there exists a sequence τ_k of positive real numbers such that τ_k converges to 0 and $\phi^{\tau_k}(p) = p$ for each natural number k. Let $t \geq 0$. By the Archimedean property of real numbers, for each natural number k there exists a natural number N_k such that

$$N_k \tau_k \leq t < (N_k + 1)\tau_k.$$

So, for each natural number k, there exists a real number r_k such that

$$t = N_k \tau_k + r_k$$

and $0 \leq r_k < \tau_k$. By the group property of flows and Proposition 2.2.5,

$$\phi^t(p) = \phi^{N_k \tau_k + r_k}(p) = \phi^{r_k}(\phi^{N_k \tau_k}(p)) = \phi^{r_k}(p)$$

for all natural numbers k. The sequence r_k converges to 0 since τ_k converges to 0 and $0 \leq r_k < \tau_k$ for all natural numbers k.

Let $A = \{r_k \mid k \in \mathbb{N}\}$. Then $0 \in \mathrm{cl}(A)$ and $\phi(A \times \{p\}) = \{\phi^t(p)\}$. Because $\{p\}$ is closed, $(0, p) \in \mathrm{cl}(A) \times \{p\} = \mathrm{cl}(A \times \{p\})$. Since ϕ is continuous and $\{\phi^t(p)\}$ is closed,

$$p = \phi(0, p) \in \mathrm{cl}(\phi(A \times \{p\})) = \mathrm{cl}(\{\phi^t(p)\}) = \{\phi^t(p)\}.$$

Hence, $\phi^t(p) = p$.

If $t < 0$, then $\phi^{-t}(p) = p$ by the preceding argument. Applying ϕ^t to both sides of this equation we obtain

$$p = \phi^t(p)$$

by Proposition 1.1.2. Thus, every periodic point of ϕ^t with period 0 is a fixed point of ϕ^t.

Example 2.2.3 showed that every fixed point of ϕ^t is a periodic point with period 0. Therefore, the set of fixed points of ϕ^t is the set of periodic points of ϕ^t with period zero. □

Notice that the hypothesis of Proposition 2.2.7 holds for any flow on a Hausdorff space.

Here are some examples of flows and their periodic points.

Example 2.2.8. Let β be a real number. Consider the matrix

$$A = \begin{pmatrix} 0 & -\beta \\ \beta & 0 \end{pmatrix}$$

and the linear differential equation

$$\dot{x} = Ax$$

on $\mathbb{R}^2$. By Example 1.1.6 the flow ϕ^t associated with this differential equation is

$$\phi^t(x) = e^{At}x.$$

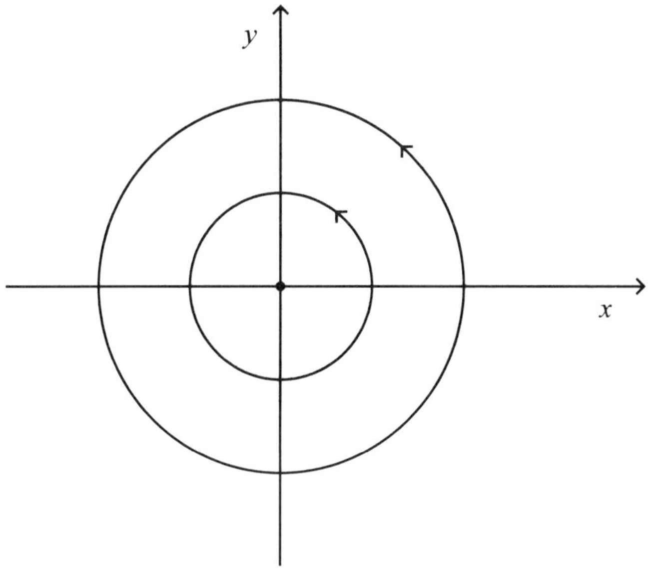

Figure 2.4. Periodic orbits of the flow of Example 2.2.8

A straightforward, though somewhat lengthy, calculation using the definition of the matrix exponential and the Maclaurin series for sine and cosine shows that

$$e^{At} = \begin{pmatrix} \cos \beta t & -\sin \beta t \\ \sin \beta t & \cos \beta t \end{pmatrix}.$$

Therefore,

$$\phi^t(x) = \begin{pmatrix} \cos \beta t & -\sin \beta t \\ \sin \beta t & \cos \beta t \end{pmatrix} x.$$

If $\beta = 0$, then every point in $\mathbb{R}^2$ is a fixed point of ϕ^t. If $\beta \neq 0$, then since $\sin \beta t$ and $\cos \beta t$ are periodic functions, and both have period $2\pi/|\beta|$, every point in $\mathbb{R}^2$ is a periodic point of ϕ^t with period $2\pi/|\beta|$. See Figure 2.4.

Example 2.2.9. Let $f : S^1 \to \mathbb{R}$ be a smooth positive function. Consider the differential equation

$$\dot{\theta} = f(\theta)$$

on S^1. We shall show that every point in S^1 is periodic with respect to ϕ^t. Regarding the circle S^1 as $\mathbb{R}/\mathbb{Z}$, define $p : \mathbb{R} \to S^1$ by

$$p(x) = x \bmod \mathbb{Z}.$$

Define $F : \mathbb{R} \to \mathbb{R}$ by

$$F = f \circ p.$$

Consider the differential equation

$$\dot{x} = F(x)$$

on $\mathbb{R}$ and its associated flow Φ^t satisfying

$$\phi^t(p(x)) = p(\Phi^t(x))$$

for all real numbers x and t.

By means of contradiction, assume that there exists $\theta \in S^1 \backslash \mathrm{Per}(\phi^t)$. Let $x \in p^{-1}(\theta)$. Since f is positive, F is positive. Thus, $\Phi^t(x)$ is an increasing function of t. We claim that $\Phi^t(x)$ is bounded above as a function of t. In particular, we shall prove that

$$\Phi^t(x) - x < 1$$

for all $t \geq 0$. By means of contradiction, assume that there exists $\tau \geq 0$ such that $\Phi^\tau(x) - x \geq 1$. Define $h : \mathbb{R} \to \mathbb{R}$ by

$$h(t) = \Phi^t(x) - x.$$

Since Φ^t is continuous, h is continuous. Furthermore, $h(0) = 0$ and $h(\tau) \geq 1$. By the Intermediate Value Theorem, there exists $T \in [0, \tau]$ such that $\Phi^T(x) - x = h(T) = 1$. Consequently, $\Phi^T(x) = x \bmod \mathbb{Z}$, and

$$\phi^T(\theta) = \phi^T(p(x)) = p(\Phi^T(x)) = p(x) = \theta.$$

Thus, θ is a periodic point of ϕ^t, contradicting the assumption that θ is not a periodic point of ϕ^t. Therefore, $\Phi^t(x) - x < 1$ for all $t \geq 0$. In particular, $\Phi^t(x)$ is bounded above as a function of t by $1 + x$. Since $\Phi^t(x)$ is increasing and bounded above as a function of t, we conclude that $\lim_{t\to\infty} \Phi^t(x)$ exists. Let

$$y = \lim_{t\to\infty} \Phi^t(x).$$

By the continuity of Φ^t and the group property of flows,

$$\begin{aligned} \Phi^s(y) &= \Phi^s\left(\lim_{t\to\infty} \Phi^t(y)\right) \\ &= \lim_{t\to\infty} \Phi^{s+t}(x) \\ &= \lim_{t\to\infty} \Phi^t(x) \\ &= y \end{aligned}$$

for all real numbers s. Hence, y is a fixed point of Φ^t. Consequently,

$$\phi^t(p(y)) = p(\Phi^t(y)) = p(y)$$

for all real numbers t. So, $p(y)$ is a fixed point of ϕ^t. However, since f is a smooth function with no zeros, the flow ϕ^t has no fixed points by Proposition 2.1.3. This is a contradiction. Therefore, every point of S^1 is a periodic point of ϕ^t. See Figure 2.5. Similarly, if $g : S^1 \to \mathbb{R}$ is a smooth

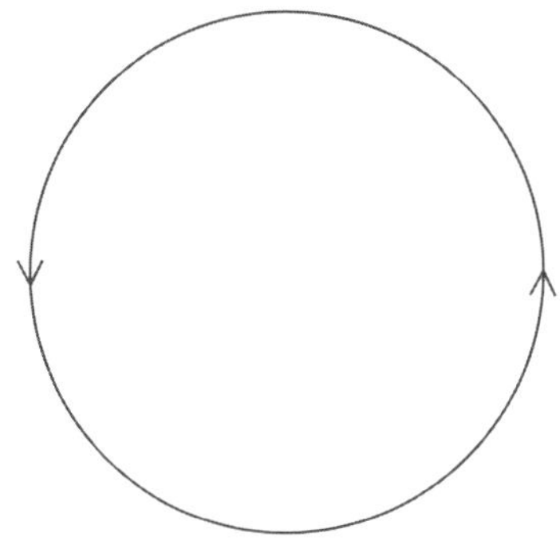

Figure 2.5. The periodic orbit of the flow of $\dot{\theta} = f(\theta)$ for $f > 0$

negative function, then every point of S^1 is periodic with respect to the flow generated by $\dot{\theta} = g(\theta)$.

In contrast to Example 2.2.9, if $f(\theta) = \sin^2 \theta$, then there are exactly two periodic points, namely the two fixed points.

Example 2.2.10. Consider the differential equation

$$\dot{\theta} = \sin^2 \theta$$

given in the angular coordinate θ on S^1. The flow ϕ^t associated with this differential equation is the arrested rotation. By Example 2.1.6 the fixed points of ϕ^t are the points with angular coordinates 0 and π. We shall show that these fixed points are the only periodic points of ϕ^t.

Regarding the circle S^1 as $\mathbb{R}/2\pi\mathbb{Z}$, define $p : \mathbb{R} \to S^1$ by

$$p(x) = x \bmod 2\pi\mathbb{Z}.$$

Consider the differential equation

$$\dot{x} = \sin^2(p(x))$$

on $\mathbb{R}$ and its corresponding flow Φ^t satisfying

$$\phi^t(p(x)) = p(\Phi^t(x))$$

for all real numbers x and t. Let θ be a periodic point of ϕ^t, and let $x \in p^{-1}(\theta)$. Since θ is periodic, there exists a positive real number τ such that $\phi^\tau(\theta) = \theta$. Thus, $\Phi^\tau(x) - x = 2\pi n$ for some integer n. Because $\dot{x} = \sin^2(p(x))$ is nonnegative and x is not a fixed point, n is positive.

The function $\Phi^t(x)$ is continuous with respect to t, $\Phi^0(x) = x$, and $\Phi^\tau(x) = x + 2\pi n$. By the Intermediate Value Theorem, $[x, x + 2\pi n] \subseteq [x, \Phi^\tau(x)]$. Thus,

$$S^1 = p([x, x + 2\pi n]) \subseteq p([x, \Phi^\tau(x)]) \subseteq \mathcal{O}^+(\theta).$$

In particular, $\mathcal{O}^+(\theta)$ contains a fixed point of ϕ^t, so that θ is a fixed point. Therefore, the fixed points of ϕ^t are the only periodic points of ϕ^t.

Example 2.2.11. Let c_1 and c_2 be real numbers. Consider the system of differential equations

$$\begin{aligned} \dot{x}_1 &= c_1, \\ \dot{x}_2 &= c_2 \end{aligned}$$

on the torus $\mathbb{T}^2$. Thinking of this system as a system of differential equations on $\mathbb{R}^2$ we obtain the flow of Example 1.1.5 with $n = 2$ and

$$c = \begin{pmatrix} c_1 \\ c_2 \end{pmatrix}$$

whose orbits are lines parallel to c for nonzero c by Example 1.1.10. Thus, regarding the torus $\mathbb{T}^2$ as $\mathbb{R}^2/\mathbb{Z}^2$, the corresponding flow $\phi^t : \mathbb{T}^2 \to \mathbb{T}^2$ is

$$\phi^t(x_1, x_2) = (x_1 + c_1 t, x_2 + c_2 t) \bmod \mathbb{Z}^2.$$

If $c_1 = c_2 = 0$, then every point of $\mathbb{T}^2$ is a fixed point of ϕ^t. If $c_1 \neq 0$ and c_2 is any real number, then the orbits of points under ϕ^t are the projections to the torus of lines in $\mathbb{R}^2$ with slope c_2/c_1. If $c_1 = 0$ and $c_2 \neq 0$, then the orbits are the projections to the torus of vertical lines in $\mathbb{R}^2$.

If c_2/c_1 is rational or $c_1 = 0$, then we refer to ϕ^t as a **rational flow on the torus.** In the case that c_2/c_1 is rational, there exist relatively prime integers p and q such that $c_2/c_1 = p/q$. Furthermore,

$$\begin{aligned} \phi^{\pm q/c_1}(x_1, x_2) &= \left(x_1 \pm c_1 \frac{q}{c_1}, x_2 \pm c_2 \frac{q}{c_1}\right) \bmod \mathbb{Z}^2 \\ &= (x_1 \pm q, x_2 \pm p) \bmod \mathbb{Z}^2 \\ &= (x_1, x_2) \bmod \mathbb{Z}^2. \end{aligned}$$

In the case that $c_1 = 0$,

$$\begin{aligned} \phi^{\pm 1/c_2}(x_1, x_2) &= \left(x_1, x_2 \pm c_2 \frac{1}{c_2}\right) \bmod \mathbb{Z}^2 \\ &= (x_1, x_2 \pm 1) \bmod \mathbb{Z}^2 \\ &= (x_1, x_2) \bmod \mathbb{Z}^2. \end{aligned}$$

Consequently, if ϕ^t is a rational flow on the torus, then every point of $\mathbb{T}^2$ is a periodic point of ϕ^t. See Figure 2.6.

We now demonstrate that if c_2/c_1 is rational and nonzero, then the period of each point is $|q/c_1| = |p/c_2|$ where p and q are relatively prime nonzero integers such that $c_2/c_1 = p/q$. We have already shown that $\phi^{\pm q/c_1}(x_1, x_2) = (x_1, x_2) \bmod \mathbb{Z}^2$. Thus,

$$\phi^{|q/c_1|}(x_1, x_2) = (x_1, x_2) \bmod \mathbb{Z}^2.$$

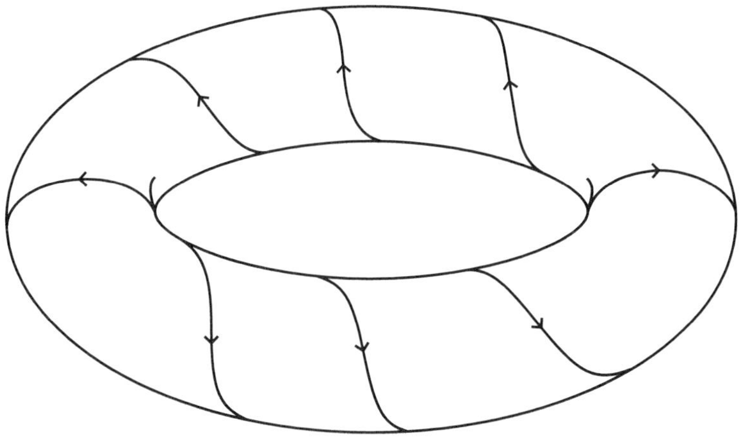

Figure 2.6. An orbit of a rational flow on the torus

It remains to show that $|q/c_1|$ is the least positive real number with this property.

If c_1 and c_2 are nonzero, τ is a positive real number, and

$$\phi^{\tau}(x_1, x_2) = (x_1 + c_1\tau, x_2 + c_2\tau) = (x_1, x_2) \bmod \mathbb{Z}^2,$$

then $c_2\tau$ is a nonzero integer, and

$$c_2\tau = \frac{c_2}{c_1}c_1\tau = \frac{p}{q}c_1\tau$$

is an integer. But, p and q are relatively prime. So, $c_1\tau = kq$ for some nonzero integer k. The fact that $\tau > 0$ implies that

$$\tau = \frac{c_1\tau}{c_1} = \frac{kq}{c_1} \geq \left|\frac{q}{c_1}\right|.$$

Therefore, the period of each point is $|q/c_1| = |p/c_2|$.

We leave it to Exercise 4 to verify that if $c_1 = 0$, then the period of each point is $1/|c_2|$, and if $c_2 = 0$ and $c_1 \neq 0$, then the period of each point is $1/|c_1|$.

Now we investigate the properties of the set of periodic points of a flow. The periodic set of a flow satisfies the Flow Invariance Property.

Proposition 2.2.12. *The periodic set of a flow is invariant with respect to the flow.*

Proof. Let ϕ^t be a flow. If $x \in \mathrm{Per}(\phi^t)$, then there exists a postive real number τ such that $\phi^{\tau}(x) = x$. The group property of flows implies that

$$\phi^{\tau}(\phi^s(x)) = \phi^{\tau+s}(x) = \phi^s(\phi^{\tau}(x)) = \phi^s(x)$$

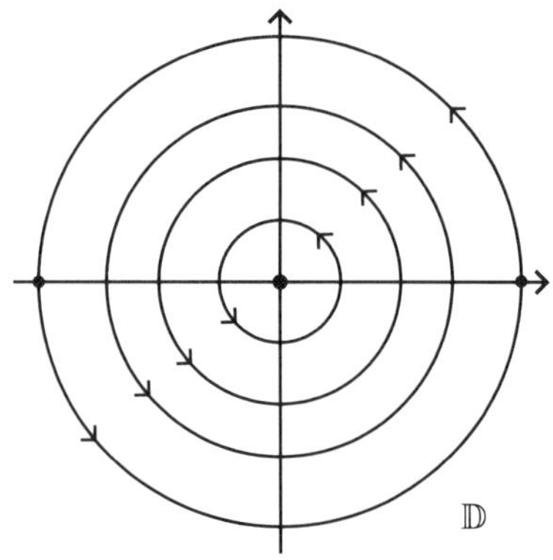

Figure 2.7. A phase portrait of the flow of Example 2.2.14

for every real number s. Thus, $\phi^s(x)$ is a periodic point of ϕ^t for every real number s. Therefore, $\mathrm{Per}(\phi^t)$ is an invariant set. □

Proposition 2.2.13. *Every periodic orbit of a flow is compact and closed.*

Proof. Let p be a periodic point of a flow ϕ^t. Let T be the period of p. Since $[0,T]$ is compact and $\phi(\cdot,p):\mathbb{R}\to X$ is continuous, $\phi([0,T],p)=\mathcal{O}(p)$ is compact. Every compact set is closed. □

The following example demonstrates that, unlike the fixed set, the periodic set of a flow does not satisfy the Closure Property. Do not confuse this fact with Proposition 2.2.13 which only asserts that each individual periodic orbit is a closed set.

Example 2.2.14. Consider the system of differential equations

$$\begin{aligned}\dot{r} &= 0,\\ \dot{\theta} &= \sin^2\theta + 1 - r^2\end{aligned}$$

in polar coordinates on the closed unit disk $\mathbb{D}$ in the complex plane. Let $\phi^t:\mathbb{D}\to\mathbb{D}$ be the flow corresponding to this system.

Since $\dot{r}=0$, the radial coordinate r is constant along each orbit. Thus, each circle centered at the origin is an invariant set with respect to ϕ^t. If $r_0\in[0,1)$, then the flow on the circle with radius r_0 centered at the origin is the flow of the differential equation

$$\dot{\theta} = \sin^2\theta + 1 - r_0^2 > 0.$$

Example 2.2.9 implies that this circle is a periodic orbit of ϕ^t. If $r_0=1$, then the invariant circle is $\partial\mathbb{D}$. The flow on $\partial\mathbb{D}$ is the arrested rotation associated with the equation

$$\dot{\theta} = \sin^2\theta.$$

By Example 2.2.10 the only periodic points of this flow are $\theta = 0$ and $\theta = \pi$. Therefore,

$$\mathrm{Per}(\phi^t) = \mathrm{int}(\mathbb{D}) \cup \{(1,0),(1,\pi)\}$$

which is not a closed subset of $\mathbb{D}$.

Topological equivalences carry periodic points to periodic points. In particular, the periodic set of a flow satisfies the Topological Invariance Property.

Proposition 2.2.15. *Let X and Y be topological spaces. If (h,ρ) is a topological equivalence from a flow $\phi^t : X \to X$ to a flow $\psi^t : Y \to Y$, then*

$$Per(\psi^t) = h(Per(\phi^t)).$$

Proof. If $x \in \mathrm{Per}(\phi^t)$, then there exists a positive real number τ such that $\phi^\tau(x) = x$. By the Archimedean Property of the real numbers, there exists a positive integer n such that $n\tau > \rho(0,x)$. Since $\rho(\cdot,x)$ is increasing and surjective, there exists a positive real number σ such that $\rho(\sigma,x) = n\tau$. By the hypothesis that (h,ρ) is a topological equivalence and Proposition 2.2.5,

$$\psi^\sigma(h(x)) = h(\phi^{\rho(\sigma,x)}(x)) = h(\phi^{n\tau}(x)) = h(x).$$

Therefore, $h(x) \in \mathrm{Per}(\psi^t)$.

If $y \in \mathrm{Per}(\psi^t)$, then a similar argument using the topological equivalence $(h^{-1},\tilde{\rho})$ from ψ^t to ϕ^t shows that $h^{-1}(y) \in \mathrm{Per}(\phi^t)$. Thus, $y \in h(\mathrm{Per}(\phi^t))$. Therefore,

$$\mathrm{Per}(\psi^t) = h(\mathrm{Per}(\phi^t)).$$ □

Here is an illustration of Proposition 2.2.15.

Example 2.2.16. Consider a 2×2 matrix B with real entries and pure imaginary eigenvalues $\pm i\beta$ for some real number β. From Example 1.1.6 the linear differential equation

$$\dot{x} = Bx$$

defines a flow $\psi^t : \mathbb{R}^2 \to \mathbb{R}^2$ by

$$\psi^t(x) = e^{Bt}x.$$

We will show that every point in $\mathbb{R}^2$ is periodic with respect to ψ^t.

Let

$$A = \begin{pmatrix} 0 & -\beta \\ \beta & 0 \end{pmatrix}.$$

As in Example 2.2.8 the linear differential equation

$$\dot{x} = Ax$$

defines a flow $\phi^t : \mathbb{R}^2 \to \mathbb{R}^2$ by

$$\phi^t(x) = e^{At}x = \begin{pmatrix} \cos\beta t & -\sin\beta t \\ \sin\beta t & \cos\beta t \end{pmatrix} x.$$

The eigenvalues of A are $\pm i\beta$, and the eigenvalues of B are also $\pm i\beta$. Thus, A and B are similar matrices. By Example 1.2.3 the flows ψ^t and ϕ^t are topologically conjugate via a homeomorphism $L : \mathbb{R}^2 \to \mathbb{R}^2$. According to Example 2.2.8, every point in $\mathbb{R}^2$ is periodic with respect to ϕ^t. Therefore, Proposition 2.2.15 and the surjectivity of L imply that

$$\text{Per}(\psi^t) = L(\text{Per}(\phi^t)) = L(\mathbb{R}^2) = \mathbb{R}^2.$$

Example 2.3.9 will show that the periodic set does not satisfy the Accumulation Property. However, demonstrating that the periodic set satisfies the Restriction Property, and the Decomposition Property is straightforward. We include statements and proofs in the interest of completeness.

Proposition 2.2.17. *If ϕ^t is a flow, then $Per(\phi^t|_{Per(\phi^t)}) = Per(\phi^t)$.*

Proof. If $p \in \text{Per}(\phi^t|_{\text{Per}(\phi^t)})$, then $p \in \text{Per}(\phi^t)$. If $p \in \text{Per}(\phi^t)$, then there exists a positive real number τ such that $\phi^\tau(p) = p$. Consequently,

$$\phi^\tau|_{\text{Per}(\phi^t)}(p) = \phi^\tau(p) = p.$$

So, $p \in \text{Per}(\phi^t|_{\text{Per}(\phi^t)})$. Therefore, $\text{Per}(\phi^t|_{\text{Per}(\phi^t)}) = \text{Per}(\phi^t)$. □

Proposition 2.2.18. *The periodic set of a flow is a union of disjoint closed sets which are invariant with respect to the flow.*

Proof. Let $\mathcal{P}$ be the collection of periodic orbits of a flow ϕ^t. Distinct orbits are disjoint by Proposition 1.1.12. Each periodic orbit is closed by Proposition 2.2.13. Every orbit of ϕ^t is invariant with respect to ϕ^t. Therefore,

$$\text{Per}(\phi^t) = \bigcup_{\mathcal{O} \in \mathcal{P}} \mathcal{O}$$

expresses $\text{Per}(\phi^t)$ as a union of disjoint closed invariant sets. □

2.3. Limit Sets

A major objective in the study of dynamical systems is to describe the eventual behavior of a flow. Birkhoff [**7**] introduced the α- and ω-limit sets of a point x with respect to a flow ϕ^t to capture the eventual behavior of $\phi^t(x)$ in forward time and backward time, respectively. Finding the α- and ω-limit sets of each point with respect to a flow ϕ^t enables us to construct a phase portrait of ϕ^t.

Definition 2.3.1. Let ϕ^t be a flow on a Hausdorff topological space X. A point $y \in X$ is an **ω-limit point** for $x \in X$ with respect to ϕ^t if there exists a sequence t_n of real numbers such that $t_n \to \infty$ as $n \to \infty$ and

$$y = \lim_{n\to\infty} \phi^{t_n}(x).$$

The **ω-limit set** of $x \in X$ with respect to ϕ^t, denoted $\omega(x)$, is the set of all ω-limit points for x with respect to ϕ^t.

A point $y \in X$ is an **α-limit point** for $x \in X$ with respect to ϕ^t if there exists a sequence t_n of real numbers such that $t_n \to \infty$ as $n \to \infty$ and

$$y = \lim_{n\to\infty} \phi^{-t_n}(x).$$

The **α-limit set** of $x \in X$ with respect to ϕ^t, denoted $\alpha(x)$, is the set of all α-limit points for x with respect to ϕ^t.

Every ω-limit point for x is a limit point of the forward orbit of x and of the orbit of x. Similarly, every α-limit point for x is a limit point of the backward orbit of x and of the orbit of x.

Let ϕ^t be a flow on a Hausdorff space X. The concepts of α- and ω-limit sets permit a reformulation of the Accumulation Property for a set $R(\phi^t)$.

Accumulation Property. If $x \in X$ and there exists a compact subset of X containing the orbit of x, then

$$\alpha(x) \cup \omega(x) \subseteq R(\phi^t).$$

The following example shows that if p is a periodic point, then $\alpha(p) = \omega(p) = \mathcal{O}(p)$.

Example 2.3.2. Let p be a periodic point of a flow ϕ^t. Since $\mathcal{O}(p)$ is closed by Proposition 2.2.13, $\mathcal{O}(p)$ contains all of its limit points. Thus, $\alpha(p) \subseteq \mathcal{O}(p)$, and $\omega(p) \subseteq \mathcal{O}(p)$.

On the other hand, if $x \in \mathcal{O}(p)$, then there exists a real number τ such that $x = \phi^\tau(p)$. If T is the period of p, then

$$\lim_{n\to\infty} \phi^{\tau \pm nT}(p) = \lim_{n\to\infty} \phi^\tau(\phi^{\pm nT}(p)) = \phi^\tau(p) = x$$

by the group property and Proposition 2.2.5. Thus, $x \in \alpha(p)$, and $x \in \omega(p)$. Consequently, $\mathcal{O}(p) \subseteq \alpha(p)$ and $\mathcal{O}(p) \subseteq \omega(p)$.

Therefore, if p is a periodic point of ϕ^t, then

$$\alpha(p) = \omega(p) = \mathcal{O}(p).$$

In particular, if p is a fixed point of ϕ^t, then $\alpha(p) = \omega(p) = \{p\}$.

Before offering further examples of α- and ω-limit sets it is advantageous to describe and classify the limit sets of flows on closed intervals of real numbers.

Proposition 2.3.3. *Let ϕ^t be a flow on a closed interval $I \subseteq \mathbb{R}$, and let $x \in I$. If $Fix(\phi^t) \cap (-\infty, x]$ is nonempty, then let*

$$p_1 = \sup(Fix(\phi^t) \cap (-\infty, x]).$$

If $Fix(\phi^t) \cap [x, \infty)$ is nonempty, then let

$$p_2 = \inf(Fix(\phi^t) \cap [x, \infty)).$$

The following statements are true.

(i) *For each real number t the function ϕ^t preserves orientation. That is, if $y, z \in I$ and $y < z$, then $\phi^t(y) < \phi^t(z)$ for all $t \in \mathbb{R}$.*

(ii) *Every periodic point of ϕ^t is a fixed point of ϕ^t.*

(iii) *The function $\phi(\cdot, x) : \mathbb{R} \to I$ is strictly monotonic or constant.*

(iv) *Let $\mathcal{O}^+(x)$ be bounded. If $\phi^t(x)$ is strictly increasing with respect to t, then $\omega(x) = \{p_2\}$ and $\mathcal{O}^+(x) = [x, p_2)$. If $\phi^t(x)$ is strictly decreasing with respect to t, then $\omega(x) = \{p_1\}$ and $\mathcal{O}^+(x) = (p_1, x]$.*

(v) *Let $\mathcal{O}^-(x)$ be bounded. If $\phi^t(x)$ is strictly increasing with respect to t, then $\alpha(x) = \{p_1\}$ and $\mathcal{O}^-(x) = (p_1, x]$. If $\phi^t(x)$ is strictly decreasing with respect to t, then $\alpha(x) = \{p_2\}$ and $\mathcal{O}^-(x) = [x, p_2)$.*

(vi) *If $\mathcal{O}^+(x)$ is unbounded, then $\omega(x)$ is empty.*

(vii) *If $\mathcal{O}^-(x)$ is unbounded, then $\alpha(x)$ is empty.*

Proof. (i) Let $y, z \in I$ such that $y < z$. Since ϕ is continuous, the function $f : \mathbb{R} \to \mathbb{R}$ defined by

$$f(t) = \phi^t(y) - \phi^t(z)$$

is continuous. Because $y < z$ and ϕ^t is injective for each $t \in \mathbb{R}$, we have $f(0) = y - z < 0$ and $f(t) \neq 0$ for all $t \in \mathbb{R}$. The continuity of f implies that $\phi^t(y) - \phi^t(z) = f(t) < 0$ for all $t \in \mathbb{R}$. Therefore, $\phi^t(y) < \phi^t(z)$ for all $t \in \mathbb{R}$. Hence, ϕ^t preserves orientation.

(ii) Let $x \in I \setminus \mathrm{Fix}(\phi^t)$. Define $f : \mathbb{R} \to \mathbb{R}$ by

$$f(t) = \phi^t(x) - x.$$

Then $f(0) = 0$. Since x is not a fixed point of ϕ^t the function f is strictly increasing by part (i). So, there does not exist a positive real number τ such that $\phi^\tau(x) - x = f(\tau) = 0$. Hence, x is not a periodic point of ϕ^t. Therefore, every periodic point of ϕ^t is a fixed point of ϕ^t.

(iii) Assume that $\phi(\cdot, x)$ is not strictly monotonic. Then there exist real numbers a, b and c such that $a < b < c$ and

$$\phi^c(x) \in [\min\{\phi^a(x), \phi^b(x)\}, \max\{\phi^a(x), \phi^b(x)\}].$$

The continuity of $\phi(\cdot, x)$, and the Intermediate Value Theorem guarantee that there exist real numbers s and t such that $a \leq s < t \leq b$ and $\phi^s(x) = \phi^t(x)$. Consequently, $\phi^{t-s}(x) = x$, so that x is a periodic point of ϕ^t. By part (ii) every periodic point of ϕ^t is a fixed point. Thus, x is a fixed point of ϕ^t, and $\phi(\cdot, x)$ is constant. Therefore, $\phi(\cdot, x) : \mathbb{R} \to I$ is strictly monotonic or constant.

(iv) If $\mathcal{O}^+(x)$ is bounded, then by part (iii) the function $\phi^t(x)$ is a bounded monotonic function of t for $t \geq 0$. Thus, $\lim_{t\to\infty} \phi^t(x)$ exists. Since I is closed, $\lim_{t\to\infty} \phi^t(x) \in I$. If t_n is a sequence of real numbers such that $t_n \to \infty$ as $n \to \infty$, then $\lim_{n\to\infty} \phi^{t_n}(x) = \lim_{t\to\infty} \phi^t(x)$. Therefore, $\omega(x) = \{\lim_{t\to\infty} \phi^t(x)\}$.

By the continuity of ϕ^t, the group property of flows, and the existence of $\lim_{t\to\infty} \phi^t(x)$,

$$\phi^s\left(\lim_{t\to\infty} \phi^t(x)\right) = \lim_{t\to\infty} \phi^s(\phi^t(x)) = \lim_{t\to\infty} \phi^{s+t}(x) = \lim_{t\to\infty} \phi^t(x)$$

for every real number s. Therefore, $\lim_{t\to\infty} \phi^t(x)$ is a fixed point of ϕ^t. Let $p = \lim_{t\to\infty} \phi^t(x)$, and assume that $\phi^t(x)$ is strictly increasing with respect to t. In this case, x is not a fixed point of ϕ^t, $\mathcal{O}^+(x) = [x, p)$ and $p \in \mathrm{Fix}(\phi^t) \cap [x, \infty)$. We shall prove that $p = p_2$.

By means of contradiction assume that $p \neq p_2$. Then there exists $q \in \mathrm{Fix}(\phi^t) \cap [x, p)$. By the Intermediate Value Theorem, $q \in \mathcal{O}(x)$. By Proposition 1.1.12, $\mathcal{O}(x) = \mathcal{O}(q)$. Since q is a fixed point of ϕ^t, the point x is a fixed point of ϕ^t, contradicting the fact that x is not a fixed point of ϕ^t. Thus, $p = p_2$. Therefore, $\omega(x) = \{p_2\}$ and $\mathcal{O}^+(x) = [x, p_2)$.

A similar argument shows that if $\phi^t(x)$ is strictly decreasing with respect to t, then $\mathcal{O}^+(x) = [x, p_2)$ and $\omega(x) = \{p_1\}$.

(v) The proof is similar to the proof of part (iv).

(vi) If $\mathcal{O}^+(x)$ is unbounded, then for every sequence t_n of real numbers such that $t_n \to \infty$ as $n \to \infty$ the limit $\lim_{n\to\infty} \phi^{t_n}(x)$ does not exist by the monotonicity of $\phi^t(x)$ with respect to t. Therefore, $\omega(x)$ is empty.

(vii) The proof is similar to the proof of part (vi). □

If the orbit of a point with respect to a flow on an interval is bounded, then some compact set contains the orbit. Thus, Proposition 2.3.3 (iii)

Figure 2.8. A phase portrait of the flow of the logistic equation

and (iv) establish the Accumulation Property for the fixed set with respect to flows on closed intervals of real numbers. Therefore, the Accumulation Property holds for the periodic set of a flow on a closed interval by Proposition 2.3.3 (ii).

Our next two examples apply Proposition 2.3.3 to the problem of determining the limit sets for points with respect to two important flows.

Example 2.3.4. Let $\phi^t : [0,1] \to [0,1]$ be the flow associated to the logistic differential equation

$$\dot{x} = x(1-x)$$

on $[0,1]$. Determining the α- and ω-limit sets for each point in $[0,1]$ with respect to ϕ^t is simple using the explicit expression for $\phi^t(x)$ given in Example 1.1.7. See Exercise 9. However, since the flow of a differential equation does not typically admit an explicit expression, we illustrate a qualitative technique for determining α- and ω-limit sets.

According to Example 2.1.5, the fixed points of ϕ^t are 0 and 1. By Example 2.3.2, $\alpha(0) = \omega(0) = \{0\}$, and $\alpha(1) = \omega(1) = \{1\}$. If $p \in (0,1)$, then neither $\mathcal{O}^+(p)$ nor $\mathcal{O}^-(p)$ contains a fixed point of ϕ^t by Proposition 1.1.12. Since $\dot{x} = x(1-x) > 0$ for $x \in (0,1)$, the function $\phi^t(p)$ increases as a function of t. Therefore, $\omega(p) = \{1\}$ and $\alpha(p) = \{0\}$ by Proposition 2.3.3. See Figure 2.8.

Example 2.3.5. Consider the differential equation

$$\dot{\theta} = \sin^2\theta$$

in the angular coordinate θ on S^1. The flow ϕ^t of this differential equation is the arrested rotation. By Example 2.1.6 the fixed points of ϕ^t are the points with angular coordinates 0 and π. By Example 2.3.2, $\alpha(0) = \omega(0) = \{0\}$ and $\alpha(\pi) = \omega(\pi) = \{\pi\}$.

Regarding the circle S^1 as $\mathbb{R}/2\pi\mathbb{Z}$, define $p : \mathbb{R} \to S^1$ by

$$p(x) = x \bmod 2\pi\mathbb{Z}.$$

Consider the differential equation

$$\dot{x} = \sin^2(p(x))$$

on $\mathbb{R}$ and its associated flow Φ^t satisfying

$$\phi^t(p(x)) = p(\Phi^t(x))$$

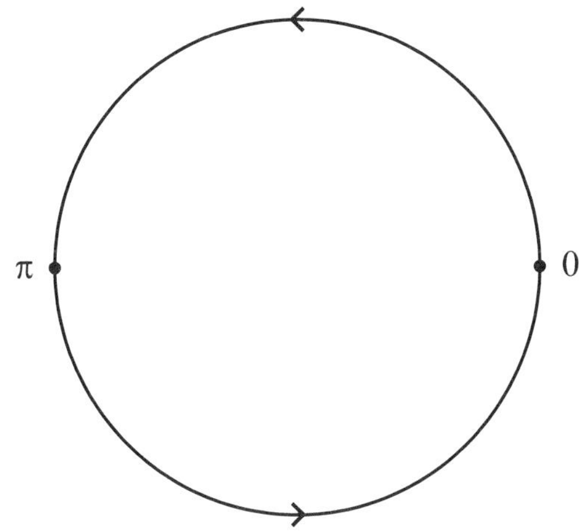

Figure 2.9. A phase portrait of the arrested rotation

for all real numbers x and t.

If $\theta = p(x)$ for some $x \in (0, \pi)$, then the orbit of x with respect to Φ^t is bounded by 0 and π. Since $\dot{x} > 0$ on $(0, \pi)$, the function $\Phi^t(x)$ is increasing with respect to t. Thus, $\alpha(x) = \{0\}$ and $\omega(x) = \{\pi\}$ with respect to Φ^t by Proposition 2.3.3. Therefore, $\alpha(\theta) = \{0\}$ and $\omega(\theta) = \{\pi\}$ by the continuity of p. If $\theta = p(x)$ for some $x \in (\pi, 2\pi)$, then $\alpha(\theta) = \{\pi\}$ and $\omega(\theta) = \{0\}$ by a similar argument. See Figure 2.9.

Before proceeding with further examples, it is useful to enumerate some important properties of α- and ω-limit sets. See also Exercise 10. Proposition 2.3.6 (viii) uses the concept of distance between a point and a nonempty closed set. If (X, d) is a metric space and A is a nonempty closed subset of X, then define

$$d(x, A) = \inf\{d(x, a) \mid a \in A\}$$

for each $x \in X$.

Proposition 2.3.6. *Let ϕ^t be a flow on a Hausdorff topological space X. Let $x \in X$.*

(i) *If $y \in \mathcal{O}(x)$, then $\omega(y) = \omega(x)$, and $\alpha(y) = \alpha(x)$.*

(ii) *If $A \subseteq X$ is closed and positively invariant, and $x \in A$, then $\omega(x) \subseteq A$. If $A \subseteq X$ is closed and negatively invariant, and $x \in A$, then $\alpha(x) \subseteq A$.*

(iii)

$$\omega(x) \subseteq \bigcap_{T \geq 0} cl\big(\{\phi^t(x) \mid t \geq T\}\big) \text{ and } \alpha(x) \subseteq \bigcap_{T \geq 0} cl\big(\{\phi^{-t}(x) \mid t \geq T\}\big)$$

with equality in both if X is first-countable.

(iv) *If X is first-countable, then $\omega(x)$ and $\alpha(x)$ are closed.*

(v) *If X is first-countable, then $\omega(x)$ and $\alpha(x)$ are invariant.*

(vi) *Let X be first-countable. If $\mathcal{O}^+(x)$ is contained in a compact subset of X, then $\omega(x)$ is nonempty, compact and connected. If $\mathcal{O}^-(x)$ is contained in a compact subset of X, then $\alpha(x)$ is nonempty, compact and connected.*

(vii) *Let X be first-countable. If $y \in \omega(x)$, then $\omega(y) \subseteq \omega(x)$, and $\alpha(y) \subseteq \omega(x)$. If $y \in \alpha(x)$, then $\alpha(y) \subseteq \alpha(x)$ and $\omega(y) \subseteq \alpha(x)$.*

(viii) *Let (X, d) be a metric space. If $\mathcal{O}^+(x)$ is contained in a compact subset of X, then*

$$\lim_{t\to\infty} d(\phi^t(x), \omega(x)) = 0.$$

If $\mathcal{O}^-(x)$ is contained in a compact subset of X, then

$$\lim_{t\to\infty} d(\phi^{-t}(x), \alpha(x)) = 0.$$

Proof. (i) If $y \in \mathcal{O}(x)$, then $y = \phi^\tau(x)$ for some real number τ. If $z \in \omega(x)$, then there exists a sequence t_n of real numbers such that $t_n \to \infty$ as $n \to \infty$ and

$$z = \lim_{n\to\infty} \phi^{t_n}(x) = \lim_{n\to\infty} \phi^{t_n}(\phi^{-\tau}(y)) = \lim_{n\to\infty} \phi^{t_n-\tau}(y).$$

Since $t_n - \tau \to \infty$ as $n \to \infty$, we obtain $z \in \omega(y)$.

On the other hand, if $z \in \omega(y)$, then there exists a sequence t_n of real numbers such that $t_n \to \infty$ as $n \to \infty$ and

$$z = \lim_{n\to\infty} \phi^{t_n}(y) = \lim_{n\to\infty} \phi^{t_n}(\phi^{\tau}(x)) = \lim_{n\to\infty} \phi^{t_n+\tau}(x).$$

Since $t_n + \tau \to \infty$ as $n \to \infty$, we obtain $z \in \omega(x)$.

Therefore, $\omega(x) = \omega(y)$. The proof for $\alpha(x)$ is similar.

(ii) Let $x \in A$. Since A is positively invariant, $\mathcal{O}^+(x) \subseteq A$. Because A is closed, A contains all limit points of $\mathcal{O}^+(x)$. Therefore, $\omega(x) \subseteq A$. The proof for $\alpha(x)$ is similar.

(iii) Let $y \in \omega(x)$. By the definition of ω-limit set,

$$y \in \operatorname{cl}\left(\{\phi^t(x) \,|\, t \geq T\}\right)$$

for each $T \geq 0$. Thus,

$$\omega(x) \subseteq \bigcap_{T\geq 0} \operatorname{cl}\left(\{\phi^t(x) \,|\, t \geq T\}\right).$$

Assume that X is first-countable. Let

$$y \in \bigcap_{T\geq 0} \operatorname{cl}\left(\{\phi^t(x) \,|\, t \geq T\}\right).$$

In particular, for each positive integer n,

$$y \in \operatorname{cl}\left(\{\phi^t(x) \,|\, t \geq n\}\right).$$

Since X is first-countable, there exists a countable basis $\{U_k\}_{k=1}^{\infty}$ for the topology of X at y. For each positive integer n the set $B_n = \bigcap_{k=1}^{n} U_k$ is a neighborhood of y. So, for each positive integer n there exists a real number $t_n \geq n$ such that $\phi^{t_n}(x) \in B_n$. Clearly, $t_n \to \infty$ as $n \to \infty$. We will show that $y = \lim_{n\to\infty} \phi^{t_n}(x)$.

Let V be a neighborhood of y. Since $\{U_k\}_{k=1}^{\infty}$ is a countable basis for the topology of X at y, there exists a positive integer N such that $U_N \subseteq V$. Thus,

$$\phi^{t_n}(x) \in B_n \subseteq B_N \subseteq U_N \subseteq V$$

for all $n \geq N$, so that $y = \lim_{n\to\infty} \phi^{t_n}(x)$. Thus, $y \in \omega(x)$.

Therefore, if X is first-countable, then

$$\omega(x) = \bigcap_{T \geq 0} \operatorname{cl}\left(\{\phi^t(x) \,|\, t \geq T\}\right).$$

Again, the proof for the α-limit set is similar.

(iv) By part (iii), the sets $\omega(x)$ and $\alpha(x)$ are intersections of closed sets whenever X is first-countable. Therefore, $\omega(x)$ and $\alpha(x)$ are closed.

(v) If $y \in \omega(x)$, then there exists a sequence t_n of real numbers such that $t_n \to \infty$ as $n \to \infty$ and $y = \lim_{n\to\infty} \phi^{t_n}(x)$. By the first-countability of X, the continuity of ϕ^t and the group property of flows,

$$\phi^{\tau}(y) = \lim_{n\to\infty} \phi^{\tau}(\phi^{t_n}(x)) = \lim_{n\to\infty} \phi^{\tau+t_n}(x)$$

for each real number τ. Since $\tau + t_n \to \infty$ as $n \to \infty$, we have $\phi^{\tau}(y) \in \omega(x)$. Therefore, $\omega(x)$ is invariant. The proof for $\alpha(x)$ is similar.

(vi) If $\mathcal{O}^+(x)$ is contained in a compact subset of X, then $\operatorname{cl}(\{\phi^t(x) \,|\, t \geq T\})$ is compact for each $T \geq 0$. Moreover, the sets $\{\phi^t(x) \,|\, t \geq T\}$ are connected for each $T \geq 0$ since $\{\phi^t(x) \,|\, t \geq T\}$ is the image of the connected set $[T, \infty)$ under the continuous function $\phi(\cdot, x) : \mathbb{R} \to X$. By part (iii), if X is first-countable, then

$$\omega(x) = \bigcap_{T \geq 0} \operatorname{cl}\left(\{\phi^t(x) \,|\, t \geq T\}\right).$$

Thus, $\omega(x)$ is an intersection of nested nonempty compact connected sets. Therefore, $\omega(x)$ is nonempty, compact and connected. The proof for $\alpha(x)$ is similar.

(vii) Let $y \in \omega(x)$. The set $\omega(x)$ is closed and invariant by parts (iv) and (v). So, $\omega(y) \subseteq \omega(x)$ and $\alpha(y) \subseteq \omega(x)$ by part (ii). Again, the proof for $\alpha(x)$ is similar.

(viii) By means of contradiction, assume that

$$\lim_{t\to\infty} d(\phi^t(x), \omega(x)) \neq 0.$$

Then there exists $\delta > 0$, and there exists a sequence t_n of real numbers such that $t_n \to \infty$ as $n \to \infty$ and

$$d(\phi^{t_n}(x), \omega(x)) \geq \delta$$

for all positive integers n. Since $\mathcal{O}^+(x)$ is contained in a compact subset of X, the sequence $\phi^{t_n}(x)$ has a limit point $z \notin \omega(x)$. But, $z \in \omega(x)$ by the definition of ω-limit point. This is a contradiction. Therefore,

$$\lim_{t\to\infty} d(\phi^t(x), \omega(x)) = 0.$$

The proof that if $\mathcal{O}^-(x)$ is contained in a compact subset of X, then $\lim_{t\to\infty} d(\phi^{-t}(x), \alpha(x)) = 0$ is similar. □

Topological equivalences carry α- and ω-limit sets to α- and ω-limit sets.

Proposition 2.3.7. *Let X and Y be Hausdorff topological spaces. If (h, ρ) is a topological equivalence from the flow $\phi^t : X \to X$ to the flow $\psi^t : Y \to Y$, then $\omega(h(x)) = h(\omega(x))$, and $\alpha(h(x)) = h(\alpha(x))$ for all $x \in X$.*

Proof. If $x \in X$ and $z \in \omega(x)$, then there exists a sequence t_n of real numbers such that $t_n \to \infty$ as $n \to \infty$ and

$$z = \lim_{n\to\infty} \phi^{t_n}(x).$$

The surjectivity and monotonicity of $\rho(\cdot, x)$ imply that for each natural number n there exists a real number s_n such that $t_n = \rho(s_n, x)$ and $s_n \to \infty$ as $n \to \infty$. Because (h, ρ) is a topological equivalence from ϕ^t to ψ^t,

$$\begin{aligned} h(z) &= h\left(\lim_{n\to\infty} \phi^{t_n}(x)\right) \\ &= \lim_{n\to\infty} h(\phi^{t_n}(x)) \\ &= \lim_{n\to\infty} h(\phi^{\rho(s_n,x)}(x)) \\ &= \lim_{n\to\infty} \psi^{s_n}(h(x)). \end{aligned}$$

So, $h(z) \in \omega(h(x))$. Thus, $h(\omega(x)) \subseteq \omega(h(x))$.

To show the reverse inclusion, recall that $(h^{-1}, \tilde{\rho})$ is a topological equivalence from ψ^t to ϕ^t. By the previous argument, $h^{-1}(\omega(h(x))) \subseteq \omega(x)$. Thus, $\omega(h(x)) \subseteq h(\omega(x))$.

Therefore,

$$\omega(h(x)) = h(\omega(x)).$$

The proof that $\alpha(h(x)) = h(\alpha(x))$ for all $x \in X$ is similar. □

In 1912 George Birkhoff [**7**] initiated a program to classify the asymptotic behavior of flows up to topological equivalence. See Sell [**40**]. The Poincaré-Bendixson Theorem classifies limit sets for planar flows. See Bendixson [**5**] and Hale [**18**].

Propositions 2.3.6 and 2.3.7 enable us to provide further examples of α- and ω-limit sets. The following example shows that the fixed set of a flow does not satisfy the Accumulation Property.

Example 2.3.8. Consider the system of differential equations

$$\begin{aligned}\dot{r} &= r(1-r),\\ \dot{\theta} &= 1\end{aligned}$$

in polar coordinates on the closed unit disk $\mathbb{D}$ in the complex plane. The flow ϕ^t corresponding to this system is the logistic rotation-dilation of Example 1.3.9. Let $z \in \mathbb{D}$ have polar coordinates (r, θ). We shall determine $\alpha(z)$ and $\omega(z)$ with respect to ϕ^t.

Case I. $r = 0$.

If $r = 0$, then z is the origin in $\mathbb{C}$. Since $\dot{r} = 0$ at the origin, z is a fixed point of ϕ^t. By Example 2.3.2, $\alpha(z) = \omega(z) = \{z\}$. See Figure 2.10.

Case II. $r = 1$.

If $r = 1$, then z lies on the boundary $\partial\mathbb{D}$ of $\mathbb{D}$. Since $\dot{r} = 0$ when $r = 1$, the boundary of $\mathbb{D}$ contains the orbit of z. On the circle $\partial\mathbb{D}$ the system of differential equations reduces to

$$\dot{\theta} = 1.$$

By Example 2.2.9, z is a periodic point of ϕ^t with orbit $\partial\mathbb{D}$. Therefore, $\omega(z) = \alpha(z) = \partial\mathbb{D}$ by Example 2.3.2. See Figure 2.10.

Case III. $0 < r < 1$.

If $0 < r < 1$, then $\lim_{t\to-\infty} r(t) = 0$ by Example 2.3.4. Thus, $\alpha(z) = \{0\}$. Example 2.3.4 also shows that $\lim_{t\to\infty} r(t) = 1$. So, $\omega(z) \subseteq \partial\mathbb{D}$. As in Case II the boundary of $\mathbb{D}$ is a periodic orbit of ϕ^t. Thus, the only nonempty invariant subset of $\partial\mathbb{D}$ is $\partial\mathbb{D}$ itself. Since $\mathbb{D}$ is compact, Proposition 2.3.6 (v) and Proposition 2.3.6 (vi) imply that the set $\omega(z)$ is a nonempty invariant set. Therefore, $\omega(z) = \partial\mathbb{D}$. See Figure 2.10.

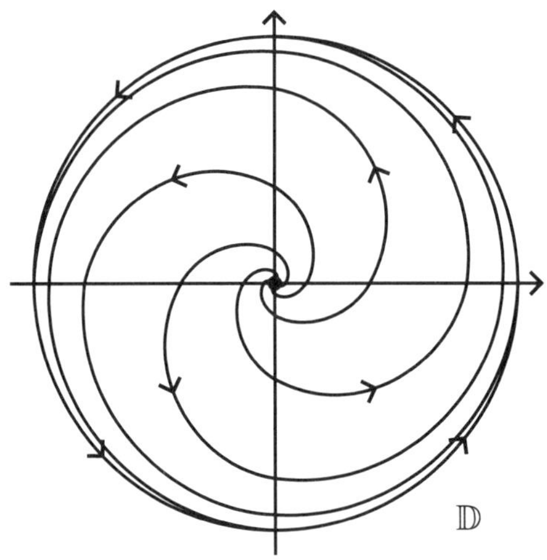

Figure 2.10. A phase portrait of the logistic rotation-dilation

The periodic set does not satisfy the Accumulation Property either.

Example 2.3.9. Consider the system of differential equations

$$\begin{aligned}\dot{r} &= r(1-r)(\sin^2\theta + 1 - r^2),\\ \dot{\theta} &= \sin^2\theta + 1 - r^2\end{aligned}$$

in polar coordinates on $\mathbb{D}$. The flow ψ^t corresponding to this system is the arrested logistic rotation-dilation of Example 1.3.9. Let $z \in \mathbb{D}$ have polar coordinates (r, θ). We will determine $\alpha(z)$ and $\omega(z)$ with respect to ψ^t.

Case I. $r = 1$.

If $r = 1$, then z lies on the boundary of $\mathbb{D}$. Since $\dot{r} = 0$ when $r = 1$, the boundary of $\mathbb{D}$ contains the orbit of z. On the circle $\partial\mathbb{D}$ the system of differential equations reduces to the arrested rotation

$$\dot{\theta} = \sin^2\theta.$$

By Example 2.1.6, the points ± 1 in the complex plane are fixed points. Applying Example 2.3.5: $\alpha(1) = \omega(1) = \{1\}$ and $\alpha(-1) = \omega(-1) = \{-1\}$; if $0 < \theta \bmod 2\pi < \pi$, then $\alpha(z) = \{1\}$ and $\omega(z) = \{-1\}$; if $\pi < \theta \bmod 2\pi < 2\pi$, then $\alpha(z) = \{-1\}$ and $\omega(z) = \{1\}$. See Figure 2.11.

Case II. $0 \leq r < 1$.

On $\text{int}(\mathbb{D})$ the flow ϕ^t of Example 2.3.8 is topologically equivalent to ψ^t by Example 1.3.9 . Extending the homeomorphism of the topological equivalence in Example 1.3.9 to the identity on $\mathbb{D}$ and applying Proposition 2.3.7,

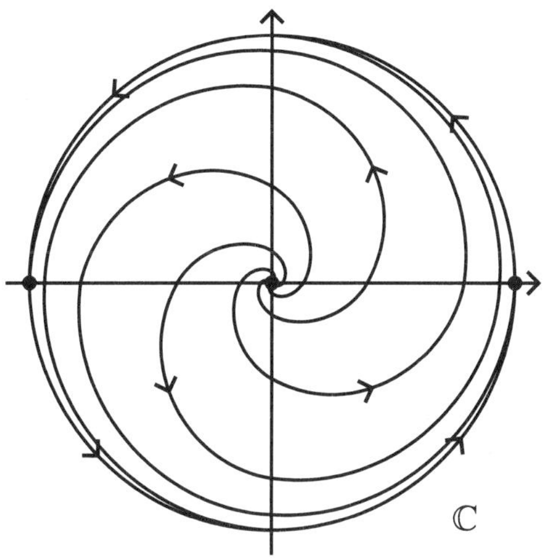

Figure 2.11. A phase portrait of the arrested logistic rotation-dilation

we obtain $\omega(z) = \partial\mathbb{D}$ if $z \neq 0$, $\omega(z) = \{0\}$ if $z = 0$, and $\alpha(z) = \{0\}$ for all $z \in \text{int}(\mathbb{D})$. See Figure 2.11.

Example 2.3.10. Let c_1 and c_2 be real numbers. Consider the system of differential equations

$$\begin{aligned} \dot{x}_1 &= c_1, \\ \dot{x}_2 &= c_2 \end{aligned}$$

on the torus $\mathbb{T}^2$. Regarding the torus as $\mathbb{R}^2/\mathbb{Z}^2$, the corresponding flow $\phi^t : \mathbb{T}^2 \to \mathbb{T}^2$ is

$$\phi^t(x_1, x_2) = (x_1 + c_1 t, x_2 + c_2 t) \text{ mod } \mathbb{Z}^2.$$

Example 2.2.11 showed that if c_2/c_1 is rational or $c_1 = 0$, then the orbit of each point of $\mathbb{T}^2$ is periodic with respect to ϕ^t. If c_2/c_1 is irrational, then we refer to ϕ^t as an **irrational flow on the torus**. See Figure 2.12. In this case, as we proceed to demonstrate, the α-limit set and ω-limit set of each point is the entire torus $\mathbb{T}^2$. Consequently, both the forward orbit and backward orbit of each point is dense in $\mathbb{T}^2$ under an irrational flow.

Assume that c_2/c_1 is irrational. Without loss of generality, we may assume that c_1 is positive. Let (p_1, p_2) represent a point in $\mathbb{T}^2$ such that $0 \leq p_1 < 1$. Consider the circle

$$C = \{(x_1, x_2) \in \mathbb{T}^2 \,|\, x_1 = p_1 \text{ mod } \mathbb{Z}\}.$$

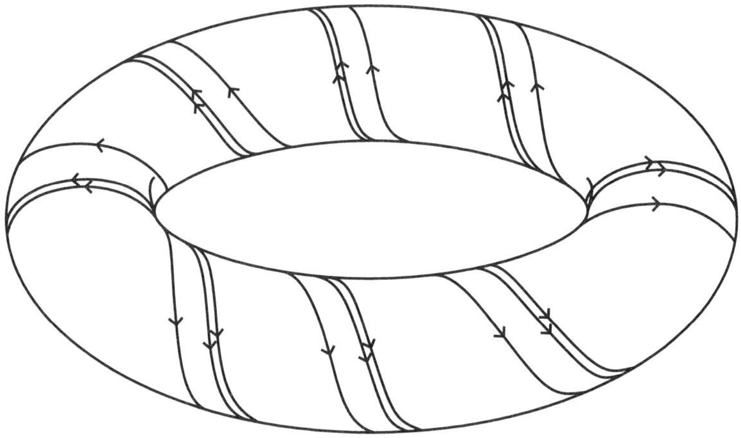

Figure 2.12. An orbit of an irrational flow on the torus

Since

$$\phi^{1/c_1}(p_1, x_2) = \left(p_1 + 1, x_2 + \frac{c_2}{c_1}\right) \bmod \mathbb{Z}^2 = \left(p_1, x_2 + \frac{c_2}{c_1}\right) \bmod \mathbb{Z}^2 \in C,$$

there is a function $\Pi : C \to C$ defined by

$$\Pi(p_1, x_2) = \phi^{1/c_1}(p_1, x_2).$$

The function Π is the rotation of the circle C by the irrational angle $2\pi c_2/c_1$.

There exists a positive integer k such that $k/c_1 \geq 1$. For each positive integer n define $\tau_n = nk/c_1$. Since

$$\phi^{\tau_n}(p_1, p_2) = (p_1 + nk, p_2 + c_2\tau_n) \bmod \mathbb{Z}^2,$$

we have $\phi^{\tau_n}(p_1, p_2) \in C$.

Let (y_1, y_2) represent a point in $\mathbb{T}^2$ such that $0 \leq y_1 < 1$. We will show that $(y_1, y_2) \in \omega(p_1, p_2)$. Define $\sigma = (p_1 - y_1)/c_1$. Since

$$\phi^{\sigma}(y_1, y_2) = (p_1, y_2 + c_2\sigma) \bmod \mathbb{Z}^2,$$

we have $\phi^{\sigma}(y_1, y_2) \in C$.

For each positive integer n let B_n denote the open ball in $\mathbb{T}^2$ of radius $1/n$ centered at (y_1, y_2). By the continuity of ϕ^{σ} the set $\phi^{\sigma}(B_n) \cap C$ is a neighborhood of $\phi^{\sigma}(y_1, y_2)$ in C. According to Theorem B.0.8 the forward orbit of $\phi^{\tau_n}(p_1, p_2)$ with respect to Π is dense in C. So, there exists a positive integer m_n such that

$$\Pi^{m_n}(\phi^{\tau_n}(p_1, p_2)) \in \phi^{\sigma}(B_n) \cap C.$$

Define a sequence of real numbers t_n by

$$t_n = \tau_n + \frac{m_n}{c_1} - \sigma.$$

Since $m_n/c_1 > 0$ and $\tau_n = nk/c_1 \geq n$, we have $t_n \geq \tau_n - \sigma \geq n - \sigma$. Thus, $t_n \to \infty$ as $n \to \infty$. We will show that $\lim_{n\to\infty} \phi^{t_n}(p_1, p_2) = (y_1, y_2)$. Let U be an open neighborhood of (y_1, y_2) in $\mathbb{T}^2$. There exists a positive integer N such that $B_n \subseteq U$ for all $n \geq N$. By the group property of flows and the definition of Π,

$$\begin{aligned}
\phi^{t_n}(p_1, p_2) &= \phi^{\tau_n + \frac{m_n}{c_1} - \sigma}(p_1, p_2) \\
&= \phi^{-\sigma}(\phi^{\frac{m_n}{c_1}}(\phi^{\tau_n}(p_1, p_2))) \\
&= \phi^{-\sigma}(\Pi^{m_n}(\phi^{\tau_n}(p_1, p_2))) \in B_n \subseteq U
\end{aligned}$$

for all $n \geq N$. Thus,

$$\lim_{n\to\infty} \phi^{t_n}(p_1, p_2) = (y_1, y_2).$$

Therefore, $\omega(p_1, p_2) = \mathbb{T}^2$ for all $(p_1, p_2) \in \mathbb{T}^2$. Similarly, $\alpha(p_1, p_2) = \mathbb{T}^2$ for all $(p_1, p_2) \in \mathbb{T}^2$. Consequently, both the forward orbit and backward orbit of each point is dense in $\mathbb{T}^2$ under an irrational flow.

Examples 2.3.9 and 2.3.10 show that neither the fixed set nor the periodic set satisfies the Accumulation Property. An α- or ω-limit set may contain points which are not periodic. Thus, in general, fixed points and periodic points do not suffice for describing the eventual behavior of a flow.

2.4. Poincaré Recurrent Points

A striking feature of an irrational flow on the torus is that both the ω-limit set and α-limit set of each point is the entire torus. In particular, for each point x the orbit of x returns near x infinitely often in both forward time and backward time. Precisely, $x \in \omega(x)$ and $x \in \alpha(x)$ for all $x \in \mathbb{T}^2$. Poincaré [**34**] identified this mode of recurrence based upon earlier work by Poisson on the stability of planetary orbits.

Definition 2.4.1. Let ϕ^t be a flow on a Hausdorff topological space X. A point $x \in X$ is **ω-recurrent** or **positively recurrent** with respect to ϕ^t if $x \in \omega(x)$ and is **α-recurrent** or **negatively recurrent** with respect to ϕ^t if $x \in \alpha(x)$. A point $x \in X$ is **Poincaré recurrent** with respect to ϕ^t if x is positively recurrent and negatively recurrent with respect to ϕ^t. The **Poincaré recurrent set** of ϕ^t, denoted $\mathrm{Rec}(\phi^t)$, is the set of Poincaré recurrent points with respect to ϕ^t.

Example 2.4.2. Let x be a periodic point. By Example 2.3.2, $x \in \mathcal{O}(x) = \omega(x) = \alpha(x)$. Therefore, every periodic point is Poincaré recurrent. In particular, every fixed point is Poincaré recurrent.

In light of Example 2.4.2, if ϕ^t is a flow on a Hausdorff topological space, then

$$\mathrm{Fix}(\phi^t) \subseteq \mathrm{Per}(\phi^t) \subseteq \mathrm{Rec}(\phi^t).$$

Example 2.4.3. Consider an irrational flow on the torus. Example 2.3.10 showed that $\omega(x) = \alpha(x) = \mathbb{T}^2$ for all $x \in \mathbb{T}^2$. So, $x \in \omega(x)$ and $x \in \alpha(x)$ for all $x \in \mathbb{T}^2$. Therefore, $\mathrm{Rec}(\phi^t) = \mathbb{T}^2$.

Example 2.4.4. Consider the differential equation

$$\dot{\theta} = \sin^2 \theta$$

in the angular coordinate on S^1. The flow ϕ^t corresponding to this differential equation is the arrested rotation. By Example 2.1.6, the fixed points of ϕ^t are the points with angular coordinates 0 and π. Example 2.3.5 showed that if $0 < \theta \bmod 2\pi < \pi$, then $\alpha(\theta) = \{0\}$ and $\omega(\theta) = \{\pi\}$, and if $\pi < \theta \bmod 2\pi < 2\pi$, then $\alpha(\theta) = \{\pi\}$ and $\omega(\theta) = \{0\}$. Therefore, $\mathrm{Rec}(\phi^t) = \{0, \pi\}$.

Now we examine the properties of flow invariance, closure and topological invariance as they relate to the set of Poincaré recurrent points.

Proposition 2.4.5. *The set of Poincaré recurrent points with respect to a flow on a Hausdorff topological space is invariant with respect to the flow.*

Proof. Let ϕ^t be a flow on a Hausdorff topological space. Let $x \in \mathrm{Rec}(\phi^t)$, and let $\tau \in \mathbb{R}$. Since x is Poincaré recurrent, $x \in \omega(x)$. So, there exists a sequence t_n of real numbers such that $t_n \to \infty$ as $n \to \infty$ and

$$x = \lim_{n\to\infty} \phi^{t_n}(x).$$

By the continuity of ϕ^t and the group property of flows,

$$\phi^\tau(x) = \phi^\tau\left(\lim_{n\to\infty} \phi^{t_n}(x)\right) = \lim_{n\to\infty} \phi^\tau(\phi^{t_n}(x)) = \lim_{n\to\infty} \phi^{t_n}(\phi^\tau(x)).$$

Thus, $\phi^\tau(x) \in \omega(\phi^\tau(x))$. Similarly, $\phi^\tau(x) \in \alpha(\phi^\tau(x))$. Thus, $\phi^\tau(x) \in \mathrm{Rec}(\phi^t)$. Therefore, $\mathrm{Rec}(\phi^t)$ is invariant with respect to ϕ^t. □

Example 2.2.14 shows that the periodic set does not satisfy the Closure Property. Because the Poincaré recurrent set coincides with the periodic set in that example, Example 2.2.14 also demonstrates that the Poincaré recurrent set does not satisfy the Closure Property. However, there is an even more interesting example of a flow such that its Poincaré recurrent set is not closed and properly contains its periodic set.

Example 2.4.6. Let ϕ^t be the flow corresponding to the system of differential equations

$$\begin{aligned}\dot{x} &= \frac{2xz}{x^2+y^2+z^2+1} - y,\\ \dot{y} &= \frac{2yz}{x^2+y^2+z^2+1} + x,\\ \dot{z} &= \frac{z^2-x^2-y^2+1}{x^2+y^2+z^2+1}\end{aligned}$$

on $\mathbb{R}^3$. We shall show that the set of Poincaré recurrent points of ϕ^t is not closed.

Define $F : \mathbb{R}^3 \to \mathbb{R}$ by

$$F(x,y,z) = \frac{4(x^2+y^2)}{(x^2+y^2+z^2+1)^2}.$$

The gradient vector field of F is

$$\nabla F(x,y,z) = \begin{pmatrix} \dfrac{8x(z^2-x^2-y^2+1)}{(x^2+y^2+z^2+1)^3} \\ \dfrac{8y(z^2-x^2-y^2+1)}{(x^2+y^2+z^2+1)^3} \\ \dfrac{-16z(x^2+y^2)}{(x^2+y^2+z^2+1)^3} \end{pmatrix}.$$

By the Chain Rule,

$$\left.\frac{d}{dt}\right|_{t=0} F(\phi^t(x,y,z)) = \nabla F(x,y,z) \cdot \left.\frac{d}{dt}\right|_{t=0} \phi^t(x,y,z) = 0$$

for all $(x,y,z) \in \mathbb{R}^3$. Since an orbit of a flow is a connected set, F is constant along each orbit of ϕ^t. So, each orbit lies within a level set of F. To determine the Poincaré recurrent set of ϕ^t we analyze the level sets of F.

Case I. $F(x,y,z) = 0$.

In this case the level set of F is the z-axis. Along the z-axis, $x = y = 0$, and the system of differential equations reduces to the constant differential equation

$$\dot{z} = 1.$$

Thus, the z-axis is an orbit of ϕ^t, and no point on the z-axis is Poincaré recurrent.

Case II. $F(x,y,z) = \dfrac{1}{r^2}$ for $r > 0$.

Parametrize the level set $F(x, y, z) = 1/r^2$ by

$$\begin{aligned} x(\alpha, \theta) &= (r + \sqrt{r^2 - 1}\cos\alpha)\cos\theta, \\ y(\alpha, \theta) &= (r + \sqrt{r^2 - 1}\cos\alpha)\sin\theta, \\ z(\alpha, \theta) &= \sqrt{r^2 - 1}\sin\alpha, \end{aligned}$$

where $\alpha \in S^1$ and $\theta \in S^1$. If $r = 1$, then this parametrization describes the unit circle in the xy-plane. If $r \neq 1$, then the coordinates α and θ describe a torus in $\mathbb{R}^3$. For each $r > 0$ the original system of differential equations reduces to the system

$$\begin{aligned} \dot{\alpha} &= -\frac{1}{r}, \\ \dot{\theta} &= 1 \end{aligned}$$

on $\mathbb{T}^2$. If r is rational, then every point on the corresponding circle or torus is periodic by Examples 2.2.9 and 2.2.11, hence Poincaré recurrent by Example 2.4.2. On the other hand, if r is irrational, then every point on the corresponding torus is Poincaré recurrent by Example 2.4.3.

Since $F(x, y, z) > 0$ for all $(x, y, z) \in \mathbb{R}^3 \setminus \{(x, y, z) \mid x = y = 0\}$, every point in $\mathbb{R}^3$ that is not on the z-axis lies on a level set of F of the type described in Case II. See Figure 2.13. Therefore,

$$\mathrm{Rec}(\phi^t) = \mathbb{R}^3 \setminus \{(x, y, z) \mid x = y = 0\},$$

and $\mathrm{Rec}(\phi^t)$ is not a closed set. Furthermore, the periodic set of ϕ^t is a nonempty proper subset of the Poincaré recurrent set of ϕ^t.

Topological equivalences carry Poincaré recurrent points to Poincaré recurrent points. In particular, the Poincaré recurrent set satisfies the Topological Invariance Property.

Proposition 2.4.7. *Let X and Y be Hausdorff topological spaces. If (h, ρ) is a topological equivalence from the flow $\phi^t : X \to X$ to the flow $\psi^t : Y \to Y$, then*

$$Rec(\psi^t) = h(Rec(\phi^t)).$$

Proof. Let $x \in \mathrm{Rec}(\phi^t)$. Since $x \in \omega(x)$, Proposition 2.3.7 implies that $h(x) \in h(\omega(x)) = \omega(h(x))$. Similarly, $h(x) \in \alpha(h(x))$. Thus, $h(x) \in \mathrm{Rec}(\psi^t)$.

Let $y \in \mathrm{Rec}(\psi^t)$. Since $y \in \omega(y)$, Proposition 2.3.7 implies that $h^{-1}(y) \in h^{-1}(\omega(y)) = \omega(h^{-1}(y))$. Similarly, $h^{-1}(y) \in \alpha(h^{-1}(y))$. So, $y \in h(\mathrm{Rec}(\phi^t))$. Therefore,

$$\mathrm{Rec}(\psi^t) = h(\mathrm{Rec}(\phi^t)). \qquad \square$$

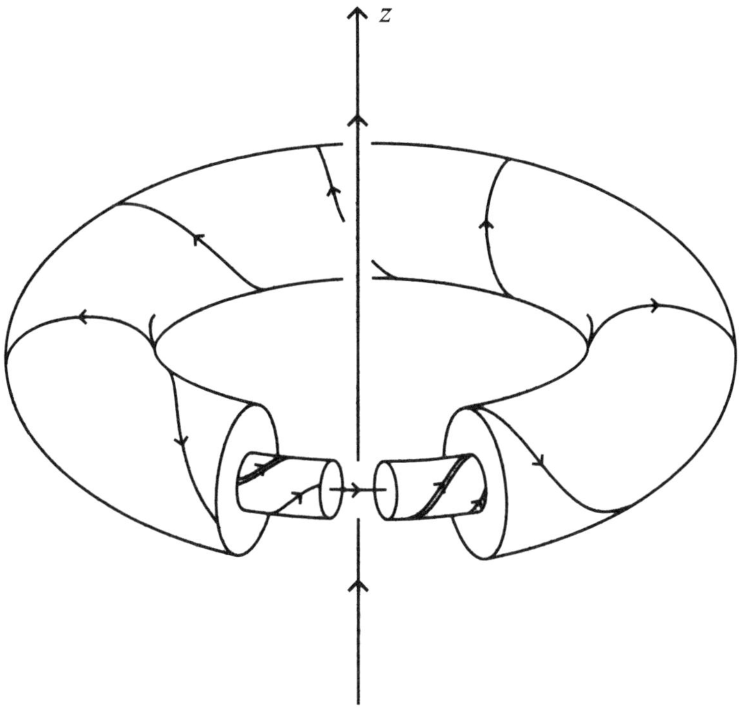

Figure 2.13. A Phase Portrait of the Flow of Example 2.4.6

In addition to failing the Closure Property, the Poincaré recurrent set has another shortcoming; it is not necessarily large enough to capture the entire eventual behavior of a flow. As the following example illustrates, the Poincaré recurrent set does not satisfy the Accumulation Property.

Example 2.4.8. Let ψ^t be the arrested logistic rotation-dilation on the closed unit disk $\mathbb{D}$ in the complex plane. By Example 2.3.9,

$$\text{Rec}(\psi^t) = \{-1, 0, 1\},$$

and $\omega(z) = \partial\mathbb{D}$ for all $z \in \text{int}(\mathbb{D}) \setminus \{0\}$. Therefore, the Poincaré recurrent set of ψ^t does not contain the ω-limit set with respect to ψ^t of every point in $\mathbb{D}$.

Finally, we establish the Restriction Property for the Poincaré recurrent set.

Proposition 2.4.9. *If ϕ^t is a flow on a Hausdorff topological space, then $Rec(\phi^t|_{Rec(\phi^t)}) = Rec(\phi^t)$.*

Proof. If $x \in \operatorname{Rec}(\phi^t|_{\operatorname{Rec}(\phi^t)})$, then $x \in \operatorname{Rec}(\phi^t)$. If $x \in \operatorname{Rec}(\phi^t)$, then $x \in \omega(x) \cap \alpha(x)$ with respect to ϕ^t. So, there exists a sequence t_n of real numbers such that $t_n \to \infty$ as $n \to \infty$ and

$$\lim_{n\to\infty} \phi^{t_n}(x) = x.$$

By the definition of the restriction of a function,

$$\lim_{n\to\infty} \phi^{t_n}|_{\operatorname{Rec}(\phi^t)}(x) = \lim_{n\to\infty} \phi^{t_n}(x) = x.$$

Thus, $x \in \omega(x)$ with respect to $\phi^t|_{\operatorname{Rec}(\phi^t)}$. Similarly, $x \in \alpha(x)$ with respect to $\phi^t|_{\operatorname{Rec}(\phi^t)}$. So, $x \in \operatorname{Rec}(\phi^t|_{\operatorname{Rec}(\phi^t)})$. Therefore,

$$\operatorname{Rec}(\phi^t|_{\operatorname{Rec}(\phi^t)}) = \operatorname{Rec}(\phi^t).$$ □

2.5. Nonwandering Points

As Examples 2.4.6 and 2.4.8 demonstrate, the Poincaré recurrent set satisfies neither the Closure Property nor the Accumulation Property. Poincaré [**34**] introduced a more general type of recurrence in the development of his famous recurrence theorem (Theorem 4.2.17). In this section we will see that this more general mode of recurrence satisfies the Closure Property and the Accumulation Property, but not the Restriction Property.

Definition 2.5.1. Let ϕ^t be a flow on a topological space X. A point $x \in X$ is **nonwandering** or **regionally recurrent** for ϕ^t if for each open neighborhood U of x there exists a real number $T > 1$ such that

$$U \cap \phi^T(U) \neq \emptyset.$$

The **nonwandering set** of ϕ^t, denoted $\Omega(\phi^t)$, is the set of nonwandering points for ϕ^t. A point which is not nonwandering for ϕ^t is **wandering** for ϕ^t. See Figure 2.14.

The modern formulation of the nonwandering set appears to originate with George Birkhoff (1884–1944) in Birkhoff [**7**]. Some older works refer to nonwandering points as regionally recurrent points since being nonwandering involves orbits of points in regions (open neighborhoods) rather than the orbit of a single point.

Example 2.5.2. If x is a fixed point for a flow ϕ^t, then x is nonwandering for ϕ^t since for each open neighborhood U of x and every $T > 1$ we have $x \in U \cap \phi^T(U)$.

Example 2.5.3. If x is a periodic point with period $T > 0$ for a flow ϕ^t, then there exists a natural number n such that $nT > 1$. Thus, x is nonwandering for ϕ^t since $\phi^{nT}(x) = x$, so that for each open neighborhood U of x we have $x \in U \cap \phi^{nT}(U)$.

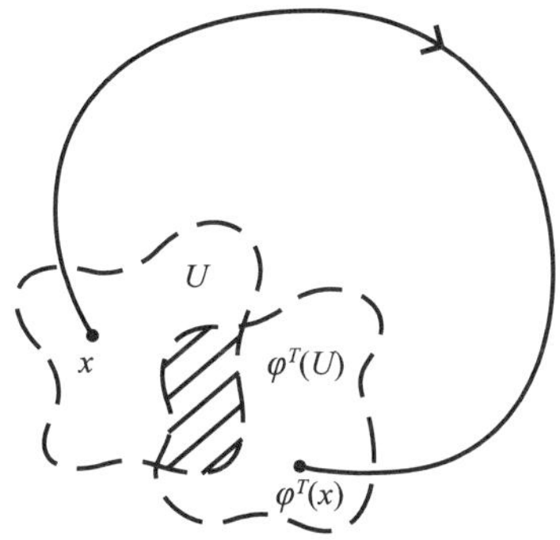

Figure 2.14. A nonwandering point

Example 2.5.4. Consider an irrational flow ϕ^t on the torus $\mathbb{T}^2$. Let $x \in \mathbb{T}^2$. By Example 2.4.3 the point x is Poincaré recurrent. So, $x \in \omega(x)$. Thus, there exists a sequence t_n of real numbers such that $t_n \to \infty$ as $n \to \infty$ and $\lim_{n\to\infty} \phi^{t_n}(x) = x$. If U is an open neighborhood of x, then since $t_n \to \infty$ as $n \to \infty$, there exists a positive integer N such that $\phi^{t_n}(x) \in U$ and $t_n > 1$ for all $n \geq N$. If $T = t_N$, then $T > 1$ and $U \cap \phi^T(U) \neq \emptyset$. Therefore, $\Omega(\phi^t) = \mathbb{T}^2$.

Corollary 2.5.15 will show that every Poincaré recurrent point is nonwandering for a flow on a Hausdorff space. Before proceeding with further examples it is useful to establish the Flow Invariance Property and the Closure Property for the nonwandering set. We shall also formulate hypotheses under which topological equivalences carry nonwandering sets to nonwandering sets.

Proposition 2.5.5. *The nonwandering set of a flow is invariant with respect to the flow.*

Proof. Let ϕ^t be a flow. Let $x \in \Omega(\phi^t)$ and $\tau \in \mathbb{R}$. Let U be an open neighborhood of $\phi^\tau(x)$. The continuity of ϕ^τ implies that $V = \phi^{-\tau}(U)$ is an open neighborhood of x. Since $x \in \Omega(\phi^t)$, there exists $T > 1$ such that $V \cap \phi^T(V) \neq \emptyset$. Consequently, $U \cap \phi^T(U) = \phi^\tau(V \cap \phi^T(V)) \neq \emptyset$. Therefore, $\Omega(\phi^t)$ is invariant. □

In contrast to the set of Poincaré recurrent points, the nonwandering set satisfies the Closure Property.

Proposition 2.5.6. *The nonwandering set of a flow is closed.*

Proof. Let ϕ^t be a flow. Let x be a limit point of $\Omega(\phi^t)$. Let U be an open neighborhood of x. Since x is a limit point of $\Omega(\phi^t)$, there exists

$y \in U \cap \Omega(\phi^t)$. The point y is nonwandering, so there exists $T > 1$ such that $U \cap \phi^T(U) \neq \emptyset$. Thus, $x \in \Omega(\phi^t)$. Therefore, $\Omega(\phi^t)$ is closed. □

We shall demonstrate that topological equivalences map nonwandering sets to nonwandering sets. Unlike the analogous situations for fixed points, periodic points and Poincaré recurrent points, the preservation of the nonwandering set under a topological equivalence holds under mild restrictions on the topology of the phase space. The weaker Topological Invariance Property does not require further assumptions; so we begin with a separate proof of this fact.

Proposition 2.5.7. *Let X and Y be topological spaces. If $h : X \to Y$ is a topological conjugacy from the flow $\phi^t : X \to X$ to the flow $\psi^t : Y \to Y$, then*

$$\Omega(\psi^t) = h(\Omega(\phi^t)).$$

Proof. Let $x \in \Omega(\phi^t)$, and let U be an open neighborhood of $h(x)$. The continuity of h implies that $h^{-1}(U)$ is an open neighborhood of x. Since x is nonwandering with respect to ϕ^t, there exists a real number $T > 1$ such that

$$h^{-1}(U) \cap \phi^T(h^{-1}(U)) \neq \emptyset.$$

Since h is a topological conjugacy,

$$\phi^T(h^{-1}(U)) = h^{-1}(\psi^T(U)).$$

So,

$$h^{-1}(U) \cap h^{-1}(\psi^T(U)) \neq \emptyset.$$

Consequently,

$$U \cap \psi^T(U) = h(h^{-1}(U) \cap h^{-1}(\psi^T(U))) \neq \emptyset.$$

Hence, $h(x) \in \Omega(\psi^t)$. Thus,

$$h\left(\Omega(\phi^t)\right) \subseteq \Omega(\psi^t).$$

A similar argument using the topological conjugacy $h^{-1} : Y \to X$ from ψ^t to ϕ^t shows that

$$\Omega(\psi^t) \subseteq h\left(\Omega(\phi^t)\right).$$

Therefore,

$$\Omega(\psi^t) = h(\Omega(\phi^t)).$$ □

Under mild restrictions on the phase spaces, topological equivalences carry nonwandering sets to nonwandering sets. In this direction, consider two lemmas.

Lemma 2.5.8. *Let ϕ^t be a flow on a first-countable Hausdorff topological space X. If x is not a periodic point of ϕ^t, then for each real number $K \geq 1$, there exists an open neighborhood V of x such that $V \cap \phi^t(V) = \emptyset$ for all $t \in [1, K]$.*

Proof. We prove the contrapositive. Let $x \in X$, and let $K \geq 1$. Assume that for every open neighborhood V of x there exists $t \in [1, K]$ such that $V \cap \phi^t(V) \neq \emptyset$. Since X is first-countable, there exists a countable base $\{V_i\}$ for the topology of X at x such that $V_1 \supseteq V_2 \supseteq V_3 \supseteq \cdots$. By our assumption, there is a sequence (x_i, t_i) in $X \times [1, K]$ with the property that $x_i \in V_i$ and $\phi^{t_i}(x_i) \in V_i \cap \phi^{t_i}(V_i)$ for each natural number i. So, the sequence x_i converges to x. The compactness of $[1, K]$ guarantees that the sequence t_i has a subsequence t_{i_j} converging to some point $\tau \in [1, K]$. By the continuity of ϕ^t, the sequence $\phi^{t_{i_j}}(x_{i_j})$ converges to $\phi^\tau(x)$. We claim that $\phi^\tau(x) = x$.

Let U be an open neighborhood of x. There exists $N > 0$ so that $V_n \subseteq U$ for each $n > N$. If $i_j > N$, then $V_{i_j} \cap \phi^{t_{i_j}}(V_{i_j}) \subseteq U$. So, if $i_j > N$, then $\phi^{t_{i_j}}(x_{i_j}) \in U$. Consequently, the sequence $\phi^{t_{i_j}}(x_{i_j})$ converges to x. Because X is Hausdorff,

$$\phi^\tau(x) = \lim_{j \to \infty} \phi^{t_{i_j}}(x_{i_j}) = x.$$

Thus, x is periodic. Therefore, by contraposition, if x is not a periodic point of ϕ^t, then for each real number $K \geq 1$, there exists a neighborhood V of x such that $V \cap \phi^t(V) = \emptyset$ for all $t \in [1, K]$. □

Lemma 2.5.9. *Let ϕ^t be a flow on a first-countable Hausdorff topological space X. If $x \in \Omega(\phi^t)$, then for each open neighborhood U of x and each real number K there exists a real number $T \geq K$ such that $U \cap \phi^T(U) \neq \emptyset$.*

Proof. We prove the contrapositive. Let $x \in X$. Assume that there exists an open neighborhood U of x and a real number K such that $U \cap \phi^T(U) = \emptyset$ for all $T \geq K$. In particular, x is not a periodic point of ϕ^t. By Lemma 2.5.8 there exists an open neighborhood V of x such that $V \cap \phi^t(V) = \emptyset$ for all $t \in [1, K]$. Consequently, $U \cap V$ is a neighborhood of x and $\phi^t(U \cap V) \cap (U \cap V) = \emptyset$ for all $t \geq 1$. Thus, x is a wandering point. Therefore, by contraposition, for each neighborhood U of x and each real number K there exists a real number $T \geq K$ such that $U \cap \phi^T(U) \neq \emptyset$. □

Proposition 2.5.10. *Let X and Y be locally compact Hausdorff topological spaces, and let (h, ρ) be a topological equivalence from the flow $\phi^t : X \to X$ to the flow $\psi^t : Y \to Y$. If $\rho : \mathbb{R} \times X \to \mathbb{R}$ and $\tilde{\rho} : \mathbb{R} \times Y \to \mathbb{R}$ are continuous, then*

$$\Omega(\psi^t) = h(\Omega(\phi^t)).$$

Proof. We begin by showing that $h(\Omega(\phi^t)) \subseteq \Omega(\psi^t)$. Let $x \in \Omega(\phi^t)$, and let V be an open neighborhood of $h(x)$. We must show that $V \cap \psi^T(V) \neq \emptyset$ for some $T > 1$.

Let $U = h^{-1}(V)$. Since X is locally compact and Hausdorff, there is an open neighborhood W of x such that $\mathrm{cl}(W)$ is compact and $\mathrm{cl}(W) \subseteq U$. Define $g : \mathrm{cl}(W) \to \mathbb{R}$ by

$$g(q) = \rho(\tilde{\rho}(1, h(q)), q).$$

Since g is continuous and $\mathrm{cl}(W)$ is compact,

$$M = \max\{g(q) \mid q \in \mathrm{cl}(W)\}$$

exists. By Lemma 2.5.9, there exists $S > M$ such that

$$W \cap \phi^S(W) \neq \emptyset.$$

Let $z \in W \cap \phi^S(W)$. Then $z = \phi^S(w)$ for some $w \in W$, and $h(z) \in V$. There exists T such that $\rho(T, w) = S$. Since $S > M$, we have $T > 1$. Applying the topological equivalence (h, ρ),

$$h(z) = h(\phi^S(w)) = h(\phi^{\rho(T,w)}(w)) = \psi^T(h(w)) \in \psi^T(V).$$

Thus,

$$h(z) \in V \cap \psi^T(V).$$

So, $V \cap \psi^T(V) \neq \emptyset$. Therefore, $h(x) \in \Omega(\psi^t)$ so that $h(\Omega(\phi^t)) \subseteq \Omega(\psi^t)$.

Similarly, using the topological equivalence $(h^{-1}, \tilde{\rho})$, we obtain $\Omega(\psi^t) \subseteq h(\Omega(\phi^t))$. Therefore, $\Omega(\psi^t) = h(\Omega(\phi^t))$. □

Let us look at three more examples of nonwandering sets.

Example 2.5.11. Consider the differential equation

$$\dot{\theta} = \sin^2 \theta$$

in the angular coordinate on S^1. The flow ϕ^t associated to this differential equation is the arrested rotation and has two fixed points with coordinates 0 and π. These fixed points are nonwandering by Example 2.5.2.

Suppose θ is not a fixed point of ϕ^t. By Example 2.3.5, the forward orbit of θ traverses S^1 counterclockwise and is bounded by one of the two fixed points. Thus, the arc $U = \{\phi^t(\theta) \mid -\frac{1}{2} < t < \frac{1}{2}\}$ on S^1 is an open neighborhood of θ with the property that $U \cap \phi^T(U) = \emptyset$ for every $T > 1$. Consequently, every point in S^1 other than 0 or π is a wandering point. Therefore, $\Omega(\phi^t) = \{0, \pi\}$.

Example 2.5.12. Consider the system of differential equations

$$\begin{aligned} \dot{r} &= r(1 - r), \\ \dot{\theta} &= 1 \end{aligned}$$

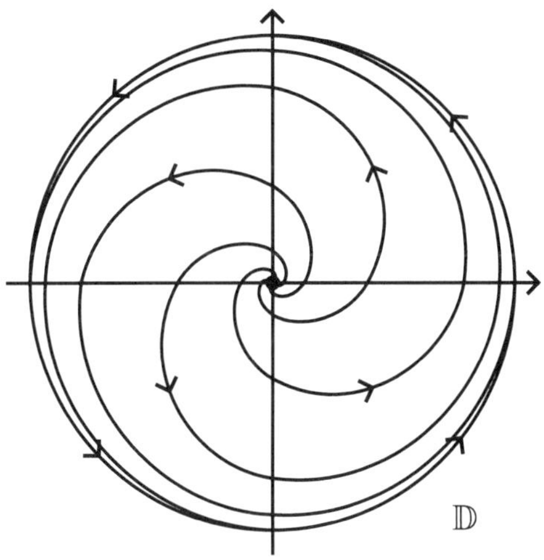

Figure 2.15. A phase portrait of the logistic rotation-dilation

in polar coordinates on the closed unit disk $\mathbb{D}$ in the complex plane. The flow ϕ^t associated to this system is the logistic rotation-dilation. See Figure 2.15. Let $z \in \mathbb{D}$ have polar coordinates (r_0, θ_0).

Case I. $r_0 = 0$.

In this case z is the origin in the complex plane, and $\dot{r} = 0$. So, z is a fixed point of ϕ^t. Hence, z is a nonwandering point of ϕ^t by Example 2.5.2.

Case II. $r_0 = 1$.

In this case $z \in \partial\mathbb{D}$. Since $\dot{r} = 0$ on $\partial\mathbb{D}$, the boundary of $\mathbb{D}$ contains the orbit of z. On $\partial\mathbb{D}$ the differential equations reduce to

$$\dot{\theta} = 1.$$

By Example 2.2.9 the point z is periodic. Thus, z is nonwandering by Example 2.5.3.

Case III. $0 < r_0 < 1$.

In this case $\dot{r} > 0$, so r is strictly increasing. If

$$U = \{(R, \theta) \mid r(-\tfrac{1}{2}) < R < r(\tfrac{1}{2})\},$$

then U is an open neighborhood of z, and

$$\phi^T(U) = \{(R, \theta) \mid r(T - \tfrac{1}{2}) < R < r(T + \tfrac{1}{2})\}.$$

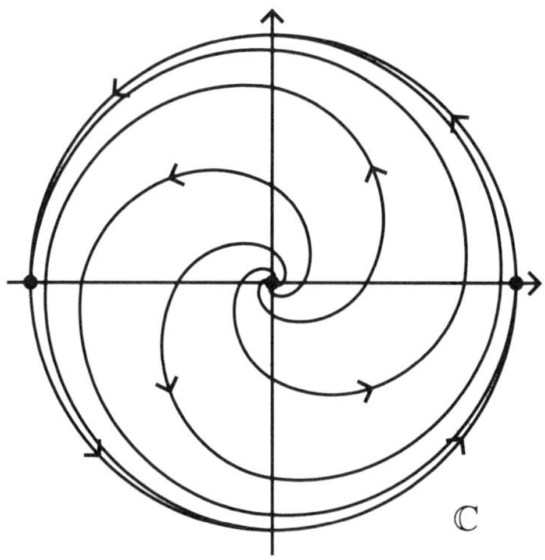

Figure 2.16. A phase portrait of the arrested logistic rotation-dilation

Since $r(\frac{1}{2}) < r(T - \frac{1}{2})$ for every $T > 1$, it follows that $U \cap \phi^T(U) = \emptyset$ for every $T > 1$. Thus, $z \notin \Omega(\phi^t)$.

Therefore,

$$\Omega(\phi^t) = \partial\mathbb{D} \cup \{0\}.$$

Example 2.5.13. Consider the system of differential equations

$$\begin{aligned}\dot{r} &= r(1-r)(\sin^2\theta + 1 - r^2),\\ \dot{\theta} &= \sin^2\theta + 1 - r^2\end{aligned}$$

in polar coordinates on the closed unit disk $\mathbb{D}$ in the complex plane. The flow ψ^t associated with this system is the arrested logistic rotation-dilation. See Figure 2.16. Let $z \in \mathbb{D}$ have polar coordinates (r_0, θ_0).

Case I. $r_0 = 1$.

In this case $z \in \partial\mathbb{D}$. Let U be an open neighborhood of z in $\mathbb{D}$, and let $w \in U \cap (\text{int}(\mathbb{D}) \setminus \{0\})$. By Example 2.3.9, $z \in \omega(w)$. Thus, there exists a sequence of real numbers t_n such that $t_n \to \infty$ as $n \to \infty$ and

$$\lim_{n\to\infty} \psi^{t_n}(w) = z.$$

So, there exists a positive integer N such that $t_n > 1$ and $\psi^{t_n}(w) \in U$ for all $n \geq N$. Consequently, if $T = t_N$, then $T > 1$ and $U \cap \psi^T(U) \neq \emptyset$. Therefore, $\partial\mathbb{D} \subseteq \Omega(\psi^t)$.

Case II. $0 \leq r_0 < 1$.

On the interior of $\mathbb{D}$ the flow ϕ^t of Example 2.5.12 is topologically equivalent via the identity homeomorphism to ψ^t by Example 1.3.9. Since the interior of $\mathbb{D}$ is locally compact and Hausdorff,

$$\Omega(\psi^t|_{\mathrm{int}(\mathbb{D})}) = \Omega(\phi^t|_{\mathrm{int}(\mathbb{D})}) = \{0\}$$

by Proposition 2.5.10.

Therefore,

$$\Omega(\psi^t) = \partial\mathbb{D} \cup \{0\}.$$

In Example 2.5.13 the nonwandering set ameliorates the shortcoming of the Poincaré recurrent set illustrated by Example 2.4.8. Unlike the fixed set, the periodic set and the Poincaré recurrent set, the nonwandering set satisfies the Accumulation Property.

Proposition 2.5.14. *If ϕ^t is a flow on a Hausdorff topological space X, then $\omega(x) \subseteq \Omega(\phi^t)$ and $\alpha(x) \subseteq \Omega(\phi^t)$ for all $x \in X$.*

Proof. Let $x \in X$ and $y \in \omega(x)$. Then there exists a sequence $t_n \to \infty$ as $n \to \infty$ such that $y = \lim_{n\to\infty} \phi^{t_n}(x)$. Let U be an open neighborhood of y in X. There exists $N > 0$ such that $\phi^{t_n}(x) \in U$ for all $n \geq N$. Since $t_n \to \infty$ as $n \to \infty$, there exists $m > N$ so that $t_m - t_N > 1$. But, $\phi^{t_m}(x) \in U$ and $\phi^{t_N}(x) \in U$. Thus,

$$\phi^{t_m}(x) = \phi^{t_m - t_N}(\phi^{t_N}(x)) \in \phi^{t_m - t_N}(U).$$

In particular, $\phi^{t_m}(x) \in U \cap \phi^{t_m - t_N}(U)$. Therefore, $y \in \Omega(\phi^t)$. The proof for $\alpha(x)$ is similar. □

Corollary 2.5.15. *The nonwandering set of a flow on a Hausdorff topological space contains the Poincaré recurrent set of the flow.*

Proof. Let ϕ^t be a flow on a Hausdorff topological space. If $x \in \mathrm{Rec}(\phi^t)$, then $x \in \omega(x) \subseteq \Omega(\phi^t)$ by Proposition 2.5.14. Therefore, $\mathrm{Rec}(\phi^t) \subseteq \Omega(\phi^t)$. □

The nonwandering set does not satisfy the Restriction Property.

Example 2.5.16. Let ψ^t be the arrested logistic rotation-dilation on the closed unit disk $\mathbb{D}$ in the complex plane. Example 2.5.13 shows that $\Omega(\psi^t) = \partial\mathbb{D} \cup \{0\}$. Since $\psi^t|_{\partial\mathbb{D}}$ is exactly the flow of Example 2.5.11 and 0 is a fixed point of ϕ^t we obtain

$$\Omega\left(\psi^t|_{\Omega(\psi^t)}\right) = \{-1, 0, 1\} \neq \Omega(\psi^t).$$

2.6. Central Points

Example 2.5.16 shows that the nonwandering set does not satisfy the Restriction Property. That is, the nonwandering set of the restriction of a flow ϕ^t to its nonwandering set may be a proper subset of the nonwandering set of ϕ^t. In this case there exists a nonwandering point x, an open neighborhood U of x, and a real number $T > 1$ such that $U \cap \phi^T(U)$ is nonempty but contains only wandering points of ϕ^t. In other words, a nonwandering point need not be nonwandering via other nonwandering points. This phenomenon is problematic if one would like to relate $\phi^T(x)$ to x through a nonwandering point in $U \cap \phi^T(U)$.

As a remedy, restrict ϕ^t to its nonwandering set, and consider the nonwandering set of the restriction. The points in the nonwandering set of the restriction are those which are nonwandering via nonwandering points of ϕ^t. By restricting many times in succession we obtain a nested collection of nonwandering sets. We shall see that the intersection of this nested collection must necessarily satisfy the Restriction Property.

To make this process precise we use ordinal numbers. For a brief introduction to ordinal numbers see Sections 17–21 of Halmos [**19**]. Let $\Omega_1(\phi^t) = \Omega(\phi^t)$, and for each positive integer n define

$$\Omega_{n+1}(\phi^t) = \Omega(\phi^t|_{\Omega_n(\phi^t)})$$

to obtain a nested collection $\{\Omega_n(\phi^t)\}$ of closed invariant sets. Denote the first infinite ordinal number by ω (not to be confused with the ω as in ω-limit set). Let

$$\Omega_\omega(\phi^t) = \bigcap_{n=1}^{\infty} \Omega_n(\phi^t).$$

If λ is a successor ordinal, that is, $\lambda = \kappa + 1$ for some ordinal κ, then define

$$\Omega_\lambda(\phi^t) = \Omega(\phi^t|_{\Omega_\kappa(\phi^t)}).$$

If λ is a limit ordinal, then define

$$\Omega_\lambda(\phi^t) = \bigcap_{\kappa<\lambda} \Omega_\kappa(\phi^t).$$

The following result due to René-Louis Baire (1874–1932) guarantees that this transfinite process stabilizes at some ordinal number.

Proposition 2.6.1. *Let X be a second-countable topological space. If*

$$\{K_1, K_2, \ldots, K_n, \ldots, K_\omega, K_{\omega+1}, \ldots\}$$

is a collection of closed subsets of X such that

$$K_1 \supseteq K_2 \supseteq \cdots \supseteq K_n \supseteq \cdots \supseteq K_\omega \supseteq K_{\omega+1} \supseteq \cdots,$$

then there exists an ordinal number β such that $K_\lambda = K_\beta$ for all ordinal numbers $\lambda \geq \beta$.

Proof. By means of contradiction assume that there does not exist an ordinal number β such that $K_\lambda = K_\beta$ for all ordinals $\lambda \geq \beta$. In this case there exists a subcollection $\{K_{\lambda_\kappa}\}$ of $\{K_\lambda\}$ such that $X \supset K_{\lambda_\kappa} \supset K_{\lambda_\mu}$ for all ordinal numbers $\kappa < \mu$. For each ordinal κ the open set $X \setminus K_{\lambda_\kappa}$ is nonempty. So, there exists $x_{\lambda_\kappa} \in X \setminus K_{\lambda_\kappa}$. Because X is second-countable there exists a countable basis $\{U_n\}$ for the topology of X. Hence, there exists a natural number n_{λ_κ} such that $x_{\lambda_\kappa} \in U_{n_{\lambda_\kappa}}$ and $U_{n_{\lambda_\kappa}} \subseteq X \setminus K_{\lambda_\kappa}$. Thus, if $\kappa < \mu$, then K_{λ_μ} does not contain $U_{n_{\lambda_\kappa}}$. Consequently, if κ and μ are distinct ordinals, then $U_{n_{\lambda_\kappa}} \neq U_{n_{\lambda_\mu}}$. Since $\{U_n\}$ is countable, the collection $\{K_{\lambda_\kappa}\}$ is countable. Because there are uncountably many ordinal numbers, this contradicts the assumption that there does not exist an ordinal number β such that $K_\lambda = K_\beta$ for all ordinals $\lambda \geq \beta$. Therefore, there exists an ordinal β such that $K_\lambda = K_\beta$ for all ordinals $\lambda \geq \beta$. □

Definition 2.6.2. Let ϕ^t be a flow on a second-countable topological space. Let $\Omega_1(\phi^t) = \Omega(\phi^t)$, and for each positive integer n define

$$\Omega_{n+1}(\phi^t) = \Omega(\phi^t|_{\Omega_n(\phi^t)})$$

to obtain a nested sequence $\{\Omega_n(\phi^t)\}$ of closed invariant sets. Let

$$\Omega_\omega(\phi^t) = \bigcap_{n=1}^{\infty} \Omega_n(\phi^t).$$

If λ is a successor ordinal, that is, $\lambda = \kappa + 1$ for some ordinal κ, then define

$$\Omega_\lambda(\phi^t) = \Omega(\phi^t|_{\Omega_\kappa(\phi^t)}).$$

If λ is a limit ordinal, then define

$$\Omega_\lambda(\phi^t) = \bigcap_{\kappa<\lambda} \Omega_\kappa(\phi^t).$$

Thus, $\{\Omega_\lambda(\phi^t)\}$ is a nested collection of closed sets. By Proposition 2.6.1 there exists a countable ordinal β such that

$$\Omega_\lambda(\phi^t) = \Omega_\beta(\phi^t)$$

for all ordinals $\lambda \geq \beta$. The set

$$C(\phi^t) = \Omega_\beta(\phi^t)$$

is the **center** of ϕ^t. An element of the center is a **central point** of ϕ^t.

Birkhoff [**7**] introduced the center. Also see Nemytskii and Stepanov [**33**].

Example 2.6.3. Every central point is nonwandering by construction.

Example 2.6.4. Let ψ^t be the arrested logistic rotation-dilation on the closed unit disk $\mathbb{D}$ in the complex plane. By Example 2.5.13,

$$\Omega_1(\psi^t) = \Omega(\psi^t) = \partial\mathbb{D} \cup \{0\}.$$

According to Example 2.5.16,

$$\Omega_2(\psi^t) = \Omega\left(\psi^t|_{\Omega_1(\psi^t)}\right) = \{-1, 0, 1\}.$$

Since -1, 0 and 1 are fixed points of ψ^t,

$$\Omega_3(\psi^t) = \Omega\left(\psi^t|_{\Omega_2(\psi^t)}\right) = \{-1, 0, 1\} = \Omega_2(\psi^t).$$

Therefore,

$$C(\psi^t) = \{-1, 0, 1\}.$$

The center of a flow is the largest subset of the phase space such that every point is nonwandering with respect to the flow in the subspace topology. Thus, the center satisfies the Restriction Property.

Proposition 2.6.5. *If ϕ^t is a flow on a second-countable topological space, then*

$$C(\phi^t|_{C(\phi^t)}) = C(\phi^t).$$

Proof. We proceed by transfinite induction. To establish the base case,

$$\Omega(\phi^t|_{C(\phi^t)}) = C(\phi^t)$$

by the definition of the center.

Let λ be an ordinal. As an inductive hypothesis, assume that

$$\Omega_\kappa(\phi^t|_{C(\phi^t)}) = C(\phi^t)$$

for all $\kappa < \lambda$. If λ is a limit ordinal, then

$$\begin{aligned}\Omega_\lambda(\phi^t|_{C(\phi^t)}) &= \bigcap_{\kappa<\lambda} \Omega_\kappa(\phi^t|_{C(\phi^t)}) \\ &= \Omega(\phi^t|_{C(\phi^t)}) \\ &= C(\phi^t).\end{aligned}$$

If λ is a successor ordinal, then $\lambda = \kappa + 1$ for some ordinal κ, and

$$\begin{aligned}\Omega_\lambda(\phi^t|_{C(\phi^t)}) &= \Omega(\phi^t|_{C(\phi^t)\cap\Omega_\kappa(\phi^t|_{C(\phi^t)})}) \\ &= \Omega(\phi^t|_{C(\phi^t)}) \\ &= C(\phi^t).\end{aligned}$$

By the Principle of Transfinite Induction, $\Omega_\lambda(\phi^t|_{C(\phi^t)}) = C(\phi^t)$ for every ordinal λ. Therefore,

$$C(\phi^t|_{C(\phi^t)}) = C(\phi^t).$$ □

The center satisfies the Flow Invariance Property and the Closure Property.

Proposition 2.6.6. *The center of a flow on a second-countable topological space is closed and invariant with respect to the flow.*

Proof. The center is an intersection of closed invariant sets. Therefore, it is closed and is invariant by Proposition 1.4.4 (ii). □

One might imagine that the center of a flow is rather small. However, the center is large enough to contain the Poincaré recurrent set. Thus, fixed points and periodic points are central as well.

Proposition 2.6.7. *The center of a flow on a second-countable Hausdorff topological space contains the Poincaré recurrent set of the flow.*

Proof. Let ϕ^t be a flow on a second-countable Hausdorff topological space X. Let $x \in \operatorname{Rec}(\phi^t)$. We shall use the Principle of Transfinite Induction to prove that $x \in \Omega_\lambda(\phi^t)$ for every ordinal λ.

Corollary 2.5.15 establishes the base case $x \in \Omega(\phi^t)$. For the inductive step, let λ be an ordinal. As an inductive hypothesis, assume that $x \in \Omega_\kappa(\phi^t)$ for all $\kappa < \lambda$. If λ is a limit ordinal, then

$$x \in \bigcap_{\kappa<\lambda} \Omega_\kappa(\phi^t) = \Omega_\lambda(\phi^t).$$

If λ is a successor ordinal, then $\lambda = \kappa + 1$ for some ordinal κ. Let V be an open neighborhood of x in the subspace topology on $\Omega_\kappa(\phi^t)$. Because V is open in $\Omega_\kappa(\phi^t)$ there exists an open set U in X such that $V = U \cap \Omega_\kappa(\phi^t)$. Since $x \in \alpha(x)$, there exists $T > 1$ such that $\phi^{-T}(x) \in U$. By the inductive hypothesis, $x \in \Omega_\kappa(\phi^t)$. The invariance of $\Omega_\kappa(\phi^t)$ implies $\phi^{-T}|_{\Omega_\kappa(\phi^t)}(x) \in V$, so that $x \in \phi^T|_{\Omega_\kappa(\phi^t)}(V)$. Thus, $x \in V \cap \phi^t|_{\Omega_\kappa(\phi^t)}(V) \neq \emptyset$. Therefore,

$$x \in \Omega\left(\phi^t|_{\Omega_\kappa(\phi^t)}\right) = \Omega_\lambda(\phi^t).$$

By the Principle of Transfinite Induction, $x \in \Omega_\lambda(\phi^t)$ for every ordinal λ, and $x \in C(\phi^t)$. Therefore,

$$\operatorname{Rec}(\phi^t) \subseteq C(\phi^t).$$ □

The center of a flow may properly contain the set of Poincaré recurrent points.

Example 2.6.8. Let ϕ^t be the flow corresponding to the system of differential equations

$$\begin{aligned}\dot{x} &= \frac{2xz}{x^2+y^2+z^2+1} - y,\\ \dot{y} &= \frac{2yz}{x^2+y^2+z^2+1} + x,\\ \dot{z} &= \frac{z^2-x^2-y^2+1}{x^2+y^2+z^2+1}\end{aligned}$$

on $\mathbb{R}^3$. Example 2.4.6 shows that

$$\operatorname{Rec}(\phi^t) = \mathbb{R}^3 \setminus \{(x,y,z) \,|\, x = y = 0\}.$$

The nonwandering set is closed by Proposition 2.5.6 and contains the Poincaré recurrent set by Proposition 2.5.15. Therefore,

$$\Omega(\phi^t) = \mathbb{R}^3.$$

Consequently, $\Omega(\phi^t|_{\Omega(\phi^t)}) = \Omega(\phi^t) = \mathbb{R}^3$, so that

$$C(\phi^t) = \mathbb{R}^3.$$

Thus, the center of ϕ^t properly contains the Poincaré recurrent set of ϕ^t.

If ϕ^t is a flow on a second-countable Hausdorff topological space, then

$$\operatorname{Fix}(\phi^t) \subseteq \operatorname{Per}(\phi^t) \subseteq \operatorname{Rec}(\phi^t) \subseteq C(\phi^t) \subseteq \Omega(\phi^t).$$

Under mild hypotheses, topological equivalences carry centers to centers. In particular, the center satisfies the Topological Invariance Property.

Proposition 2.6.9. *Let X and Y be locally compact second-countable Hausdorff topological spaces, and let (h,ρ) be a topological equivalence from the flow $\phi^t : X \to X$ to the flow $\psi^t : Y \to Y$. If $\rho : \mathbb{R}\times X \to \mathbb{R}$ and $\tilde{\rho} : \mathbb{R}\times Y \to \mathbb{R}$ are continuous, then*

$$C(\psi^t) = h(C(\phi^t)).$$

Proof. We proceed by transfinite induction. Proposition 2.5.10 establishes the base case

$$\Omega_1(\psi^t) = \Omega(\psi^t) = h(\Omega(\phi^t)) = h(\Omega_1(\phi^t)).$$

Let λ be an ordinal. As an inductive hypothesis, assume that $\Omega_\kappa(\psi^t) = h(\Omega_\kappa(\phi^t))$ for all $\kappa < \lambda$. If λ is a limit ordinal, then

$$\begin{aligned} \Omega_\lambda(\psi^t) &= \bigcap_{\kappa<\lambda} \Omega_\kappa(\psi^t) \\ &= \bigcap_{\kappa<\lambda} h(\Omega_\kappa(\phi^t)) \\ &= h\left(\bigcap_{\kappa<\lambda} \Omega_\kappa(\phi^t)\right) \\ &= h(\Omega_\lambda(\phi^t)). \end{aligned}$$

If λ is a successor ordinal, then $\lambda = \kappa + 1$ for some ordinal κ. Since (h, ρ) is a topological equivalence from ϕ^t to ψ^t, and $\Omega_\kappa(\psi^t) = h(\Omega_\kappa(\phi^t))$ by the inductive hypothesis, the pair $(h|_{\Omega_\kappa(\phi^t)}, \rho|_{\mathbb{R}\times\Omega_\kappa(\phi^t)})$ is a topological equivalence from $\phi^t|_{\Omega_\kappa(\phi^t)}$ to $\psi^t|_{\Omega_\kappa(\psi^t)}$. By Proposition 2.5.10,

$$\begin{aligned} \Omega_\lambda(\psi^t) &= \Omega(\psi^t|_{\Omega_\kappa(\psi^t)}) \\ &= h(\Omega(\phi^t|_{\Omega_\kappa(\phi^t)})) \\ &= h(\Omega_\lambda(\phi^t)). \end{aligned}$$

By the Principle of Transfinite Induction, $\Omega_\lambda(\psi^t) = h(\Omega_\lambda(\phi^t))$ for every ordinal λ. Therefore,

$$C(\psi^t) = h(C(\phi^t)).$$ □

Unfortunately, the center does not satisfy the Accumulation Property. In this respect the center shares a shortcoming of the Poincaré recurrent set.

Example 2.6.10. Let ψ^t be the arrested logistic rotation-dilation flow on the closed unit disk $\mathbb{D}$ in the complex plane. By Example 2.6.4,

$$C(\psi^t) = \{-1, 0, 1\}.$$

If $z \in \text{int}(\mathbb{D}) \setminus \{0\}$, then $\omega(z) = \partial\mathbb{D}$ by Example 2.3.9. Therefore, the center does not contain the ω-limit set of every point in $\mathbb{D}$.

2.7. Chain Recurrent Points

Conley [**11**] introduced a more general type of recurrence called chain recurrence. On compact spaces chain recurrence satisfies all of the desirable properties enumerated in the introduction to this chapter.

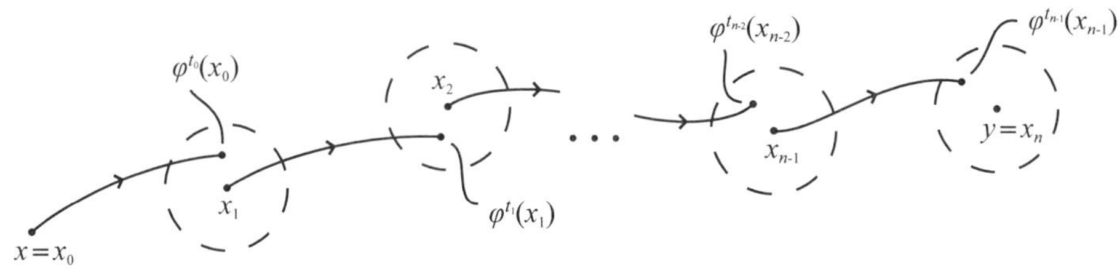

Figure 2.17. An (ϵ, T)-chain for a flow

2.7.1. The Definition of the Chain Recurrent Set.

Definition 2.7.1. Let ϕ^t be a flow on a metric space (X, d). Given $\epsilon > 0$, $T > 0$, and $x, y \in X$, an (ϵ, T)**-chain** from x to y with respect to ϕ^t and d is a pair of finite sequences $x = x_0, x_1, \ldots, x_{n-1}, x_n = y$ in X and $t_0, \ldots, t_{n-1}$ in $[T, \infty)$, denoted together by $(x_0, \ldots, x_n; t_0, \ldots, t_{n-1})$, such that

$$d(\phi^{t_i}(x_i), x_{i+1}) < \epsilon$$

for $i = 0, 1, 2, ..., n-1$. See Figure 2.17.

Definition 2.7.2. Let ϕ^t be a flow on a metric space (X, d). The **forward chain limit set** of $x \in X$ with respect to ϕ^t and d is the set

$$\Omega^+(x) = \bigcap_{\epsilon, T > 0} \{y \in X \mid \exists \text{ an } (\epsilon, T)\text{-chain from } x \text{ to } y \text{ with respect to } \phi^t\}.$$

The **backward chain limit set** of $x \in X$ with respect to ϕ^t and d is the set

$$\Omega^-(x) = \bigcap_{\epsilon, T > 0} \{y \in X \mid \exists \text{ an } (\epsilon, T)\text{-chain from } x \text{ to } y \text{ with respect to } \phi^{-t}\}.$$

Backward and forward chain limit sets are analogous to α- and ω-limit sets, respectively.

Definition 2.7.3. Let ϕ^t be a flow on a metric space (X, d). Two points $x, y \in X$ are **chain equivalent** with respect to ϕ^t and d if $y \in \Omega^+(x)$ and $x \in \Omega^+(y)$.

Two points x and y are chain equivalent if and only if for every $\epsilon > 0$ and every $T > 0$ there exists an (ϵ, T)-chain from x to y and there exists an (ϵ, T)-chain from y to x. Theorem 2.7.18 provides four equivalent formulations of chain equivalence for flows on compact spaces.

Definition 2.7.4. Let ϕ^t be a flow on a metric space (X, d). A point $x \in X$ is called **chain recurrent** with respect to ϕ^t and d if x is chain equivalent

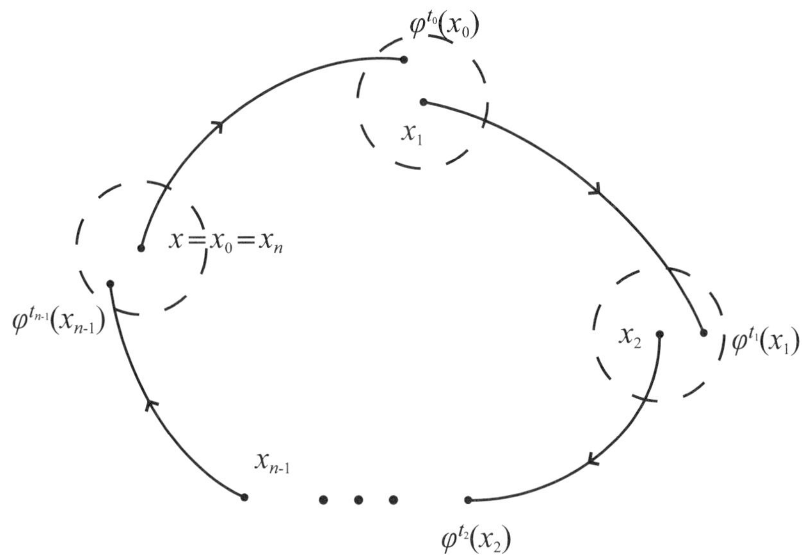

Figure 2.18. Chain recurrence

to itself. The set of all chain recurrent points of ϕ^t, denoted $\mathcal{R}(\phi^t)$, is the **chain recurrent set** of ϕ^t. See Figure 2.18.

While the language of chain limit sets is not necessary to define chain recurrence, it allows us to draw a parallel with Poincaré recurrence. A point x is chain recurrent if and only if $x \in \Omega^+(x)$. Exercise 15 asks you to show that if the phase space is compact, then $x \in \Omega^+(x)$ if and only if $x \in \Omega^-(x)$. Notice the similarity to Poincaré recurrence: A point x is Poincaré recurrent if and only if $x \in \omega(x)$ and $x \in \alpha(x)$.

Conley [**11**] defines chain recurrence for flows on compact Hausdorff topological spaces using nets. For the sake of clarity and simplicity we restrict our attention to metric spaces with an emphasis on the compact case. Hurley [**21**], [**22**] address chain recurrence on spaces which are not compact.

To imagine an (ϵ, T)-chain from x to y imagine following the orbit of x for a time at least T, then jumping a distance less than ϵ to another point, following its orbit for a time at least T, then jumping a distance less than ϵ to another point, and continuing finitely many steps in this manner until jumping onto y from a distance less than ϵ away. A point x is chain recurrent if no matter how long a time you must flow and no matter how short a distance you must jump, it is always possible to return to x. Another way to think about chain recurrence is to imagine a chain recurrent point as a point which is within ϵ of being periodic for every $\epsilon > 0$.

2.7.2. Examples of Chain Recurrent Sets.

Example 2.7.5. Every periodic point is a chain recurrent point. Let x be a periodic point with period S. Let $\epsilon > 0$ and $T > 0$. If $S > 0$, then there exists an integer n so that $nS > T$. Therefore, $(x, x; nS)$ is an (ϵ, T)-chain from x to itself. If $S = 0$, then x is a fixed point, and $(x, x; T)$ is an (ϵ, T)-chain from x to itself. Hence, $x \in \Omega^+(x)$. Therefore, x is chain recurrent. In particular, every fixed point is a chain recurrent point.

Example 2.7.6. Consider the differential equation

$$\dot{\theta} = \sin^2 \theta$$

in the angular coordinate on S^1. The flow ϕ^t associated to this differential equation is the arrested rotation. In Example 2.1.6 we saw that ϕ^t has two fixed points, 0 and π. Example 2.5.11 shows that the nonwandering set of ϕ^t is $\{0, \pi\}$. In contrast, we shall demonstrate that the chain recurrent set of ϕ^t is S^1.

Let $\theta \in S^1$. If θ is a fixed point, then θ is chain recurrent by Example 2.7.5. Denote the usual metric on S^1 by d. Let $\epsilon > 0$ and $T > 0$. If $0 < \theta \bmod 2\pi < \pi$, then $\omega(\theta) = \{\pi\}$ by Example 2.3.5. Thus, there exists $t_0 > T$ so that $d(\phi^{t_0}(\theta), \pi) < \epsilon/2$. Let θ_1 be any point in the $\epsilon/2$-ball around π such that $\pi < \theta_1 \bmod 2\pi < 2\pi$. By the triangle inequality,

$$d(\phi^{t_0}(\theta), \theta_1) < \epsilon.$$

Since $\pi < \theta_1 \bmod 2\pi < 2\pi$, Example 2.3.5 shows that $\omega(\theta_1) = \{0\}$. Thus, there exists $t_1 > T$ so that $d(\phi^{t_1}(\theta_1), 0) < \epsilon/2$. Example 2.3.5 shows that $\alpha(\theta) = \{0\}$. Thus, there exists θ_2 in the $\epsilon/2$-ball around 0 and there exists $t_2 > T$ such that $\phi^{t_2}(\theta_2) = \theta$. By the triangle inequality,

$$d(\phi^{t_1}(\theta_1), \theta_2) < \epsilon.$$

Therefore, $(\theta, \theta_1, \theta_2, \theta; t_0, t_1, t_2)$ is an (ϵ, T)-chain from θ to θ. A similar argument shows that if $\pi < \theta \bmod 2\pi < 2\pi$, then there is an (ϵ, T)-chain from θ to θ. Since ϵ and T are arbitrary positive real numbers, $\theta \in \Omega^+(\theta)$. Consequently,

$$\mathcal{R}(\phi^t) = S^1.$$

The following example illustrates a style of argument which plays an important role in the remainder of the text. We shall encounter versions of this argument again in Example 2.7.9, in the proof of Lemma 4.5.4 and in the proof of Proposition 4.5.5.

Example 2.7.7. Let a and b be real numbers with $a < b$. Let $f : [a, b] \to \mathbb{R}$ be a smooth function such that $f(a) = f(b) = 0$ and $f(x) < 0$ for all $x \in (a, b)$. We shall require this level of generality in Example 2.7.8. If you prefer a concrete example, then the function $f(x) = (x - a)(x - b)$ suffices.

Consider the flow ϕ^t associated to the differential equation

$$\dot{x} = f(x)$$

on $[a, b]$. We shall refer to this flow as the **decreasing segment flow**. By Example 2.1.3 the fixed points of ϕ^t are the zeros of f. Thus,

$$\mathrm{Fix}(\phi^t) = \{a, b\}.$$

By Example 2.7.5 fixed points are chain recurrent. So,

$$\{a, b\} \subseteq \mathcal{R}(\phi^t).$$

We shall prove that $\mathcal{R}(\phi^t) = \{a, b\}$. Notice that $\phi^t(x)$ is a strictly decreasing function of t for all $x \in (a, b)$ because $f(x) < 0$ for each $x \in (a, b)$. This insight is crucial to our argument.

Let $x \in (a, b)$ so that x is not a fixed point. Let $T > 0$. We shall produce an $\epsilon > 0$ so that there is no (ϵ, T)-chain from x to itself.

First, we claim that there exists $\delta > 0$ such that if $y < \phi^T(x) + \delta$, then

$$\phi^t(y) < \phi^T(x)$$

for all $t \geq T$. This means that if you start at a point y less than $\phi^T(x) + \delta$ and flow for at least time T, you must end up at a point less than $\phi^T(x)$. Let $\delta = x - \phi^T(x)$. Since ϕ^t is a strictly decreasing function of t and $T > 0$, we have $\delta > 0$. If $y < \phi^T(x) + \delta = x$, then

$$\phi^t(y) \leq \phi^T(y) < \phi^T(x)$$

for all $t \geq T$.

Now we shall produce an $\epsilon > 0$ so that there is no (ϵ, T)-chain from x to itself. Let $\epsilon = \min\{\delta, x - \phi^T(x)\}$. By means of contradiction, assume that there exists an (ϵ, T)-chain

$$(x = x_0, x_1, ..., x_n = x; t_0, t_1, ..., t_{n-1})$$

from x to itself. If d denotes the standard metric on $[a, b]$, then

$$|\phi^{t_0}(x_0) - x_1| = d(\phi^{t_0}(x_0), x_1) < \epsilon \leq \delta.$$

Since $t_0 \geq T$ and $\phi^t(x_0)$ is a decreasing function of t,

$$x_1 < \phi^{t_0}(x_0) + \delta \leq \phi^T(x_0) + \delta.$$

Since $t_1 \geq T$,

$$\phi^{t_1}(x_1) < \phi^T(x_0).$$

Now, $|\phi^{t_1}(x_1) - x_2| = d(\phi^{t_1}(x_1), x_2) < \epsilon \leq \delta$, so that

$$x_2 < \phi^{t_1}(x_1) + \delta < \phi^T(x_0) + \delta.$$

Thus,

$$\phi^{t_2}(x_2) < \phi^T(x_0).$$

Continuing in this manner we obtain

$$\phi^{t_{n-1}}(x_{n-1}) < \phi^T(x_0) < x_0.$$

Since $x_n = x$ and $t_{n-1} \geq T$,

$$\begin{aligned} |\phi^{t_{n-1}}(x_{n-1}) - x_n| &= |\phi^{t_{n-1}}(x_{n-1}) - x| \\ &> x - \phi^T(x). \end{aligned}$$

On the other hand, by the definition of ϵ,

$$|\phi^{t_{n-1}}(x_{n-1}) - x_n| < \epsilon \leq x - \phi^T(x).$$

This is a contradiction. Hence, $x \notin \mathcal{R}(\phi^t)$. Therefore,

$$\mathcal{R}(\phi^t) = \{a, b\}.$$

Example 2.7.8. Let $C \subset [0,1]$ be the Cantor ternary set. Let $f : [0,1] \to \mathbb{R}$ be a smooth function such that $f(x) = 0$ for all $x \in C$ and $f(x) < 0$ for all $x \in [0,1] \setminus C$.

Let ϕ^t be the flow associated to the differential equation

$$\dot{x} = f(x)$$

on $[0,1]$. By Example 2.1.3 the fixed points of ϕ^t are exactly the zeros of f. Hence, $\mathrm{Fix}(\phi^t) = C$. Since fixed points are chain recurrent, $C \subseteq \mathcal{R}(\phi^t)$. We claim that $\mathcal{R}(\phi^t) = C$.

Let $x \in [0,1] \setminus C$. By properties of the Cantor ternary set, there exist $a, b \in C$ so that $x \in [a,b]$ and $(a,b) \cap C = \emptyset$. The flow $\phi^t|_{[a,b]}$ is exactly a flow of the type studied in Example 2.7.7. That example shows that $x \notin \mathcal{R}(\phi^t)$. Therefore,

$$\mathcal{R}(\phi^t) = C.$$

Our next example illustrates a more sophisticated version of the argument given in Example 2.7.7 and builds on the analysis of the circle flow in Example 2.7.6.

Example 2.7.9. Consider the system of differential equations

$$\begin{aligned} \dot{r} &= r(1-r)(\sin^2\theta + 1 - r^2), \\ \dot{\theta} &= \sin^2\theta + 1 - r^2 \end{aligned}$$

in polar coordinates on the closed unit disk $\mathbb{D}$ in the complex plane. The flow ψ^t corresponding to this system is the arrested logistic rotation-dilation. The flow ψ^t has three fixed points: $-1, 0$ and 1. We shall demonstrate that

$$\mathcal{R}(\psi^t) = \partial\mathbb{D} \cup \{0\}.$$

Since $r = 1$ on the boundary of $\mathbb{D}$, the flow restricted to the boundary is the flow of Example 2.7.6 in which every point is chain recurrent. By Example 2.7.5, every fixed point is chain recurrent. So, 0 is chain recurrent. We must show that 0 is the only chain recurrent point in the interior of $\mathbb{D}$.

Let $z \in \operatorname{int}(\mathbb{D}) \setminus \{0\}$. For each $T > 0$ we seek an $\epsilon > 0$ so that there is no (ϵ, T)-chain from z to itself. Define $L : \mathbb{D} \to \mathbb{R}$ by

$$L(w) = 1 - |w|.$$

In fact, $L(w)$ is the distance from w to the boundary of $\mathbb{D}$. Since $0 < r < 1$ on $\operatorname{int}(\mathbb{D}) \setminus \{0\}$,

$$\dot{r} = r(1-r)(\sin^2\theta + 1 - r^2) > 0$$

on $\operatorname{int}(\mathbb{D}) \setminus \{0\}$. Consequently, $L(\psi^t(w))$ is strictly decreasing as a function of t for each $w \in \operatorname{int}(\mathbb{D}) \setminus \{0\}$.

Let $a = L(\psi^T(z))$ and $b = L(z)$. Then $0 < a < b < 1$. We claim that there exists $\delta > 0$ such that if $L(w) < a + \delta$, then

$$L(\psi^t(w)) < a$$

for all $t \geq T$. Let $\delta = 1 - a - |z|$. If $L(w) < a + \delta$, then $1 - |w| < 1 - |z|$, so that $|w| > |z|$. Because $\dot{r}$ is positive and independent of θ, we have $L(\psi^T(w)) < L(\psi^T(z))$. Since L is decreasing along the orbit of w we obtain

$$L(\psi^t(w)) \leq L(\psi^T(w)) < L(\psi^T(z)) = a$$

for all $t \geq T$.

Let $\eta = \min\{\delta, b - a\}$. Since $\mathbb{D}$ is compact and L is continuous, L is uniformly continuous. Thus, there exists $\epsilon > 0$ such that if $w_1, w_2 \in \mathbb{D}$ and $|w_1 - w_2| < \epsilon$, then

$$|L(w_1) - L(w_2)| < \eta.$$

By means of contradiction, assume that there is an (ϵ, T)-chain

$$(z = z_0, z_1, \ldots, z_{n-1}, z_n = z; t_0, \ldots, t_{n-1})$$

from z to itself. Since

$$|\psi^{t_0}(z_0) - z_1| < \epsilon,$$

the uniform continuity of L guarantees that

$$|L(\psi^{t_0}(z_0)) - L(z_1)| < \eta \leq \delta.$$

Because $t_0 \geq T$,

$$L(z_1) < L(\psi^{t_0}(z_0)) + \delta \leq L(\psi^T(z_0)) + \delta = a + \delta.$$

Thus,

$$L(\psi^{t_1}(z_1)) < a.$$

Since

$$|\psi^{t_1}(z_1) - z_2| < \epsilon,$$

the uniform continuity of L guarantees that

$$|L(\psi^{t_1}(z_1)) - L(z_2)| < \eta \leq \delta.$$

Because $t_1 \geq T$,

$$L(z_2) < L(\psi^{t_1}(z_1)) + \delta < a + \delta.$$

Thus,

$$L(\psi^{t_2}(z_2)) < a.$$

Continuing in this manner we obtain

$$L(\psi^{t_{n-1}}(z_{n-1})) < a.$$

Since $L(z_n) = L(z) = b$,

$$|L(\psi^{t_{n-1}}(z_{n-1})) - L(z_n)| = L(z) - L(\psi^{t_{n-1}}(z_{n-1})) > b - a.$$

On the other hand, since $|\psi^{t_{n-1}}(z_{n-1}) - z_n| < \epsilon$, the uniform continuity of L implies that

$$|L(\psi^{t_{n-1}}(z_{n-1})) - L(z_n)| < \eta \leq b - a.$$

This is a contradiction. Consequently, there is no (ϵ, T)-chain from z to itself. Therefore,

$$\mathcal{R}(\psi^t) = \partial\mathbb{D} \cup \{0\}.$$

The function L which plays a central role in the analysis of Example 2.7.9 is an example of a complete Lyapunov function. See Section 4.7. Among the key properties of a complete Lyapunov function are that it strictly decreases along orbits in the complement of the chain recurrent set, and that it is constant on collections of chain equivalent points. The identity function on the interval $[a, b]$ plays a similar role in Example 2.7.7. Chapter 4 investigates complete Lyapunov functions and their intimate connection to chain recurrence.

2.7.3. Elementary Properties of the Chain Recurrent Set. The chain recurrent set satisfies the Flow Invariance Property and the Closure Property. We now prove that the chain recurrent set is closed, but defer the proof of invariance to Corollary 3.3.8 where it follows easily from other considerations.

Proposition 2.7.10. *The chain recurrent set of a flow on a metric space is closed.*

Proof. Let ϕ^t be a flow on a metric space with metric d. Let y be a limit point of $\mathcal{R}(\phi^t)$. Let $\epsilon > 0$ and $T > 0$. We shall prove that $y \in \mathcal{R}(\phi^t)$ by constructing an (ϵ, T)-chain from y to itself.

By the continuity of ϕ^T, there exists $\delta > 0$ such that if $d(x,y) < \delta$, then

$$d(\phi^T(x), \phi^T(y)) < \frac{\epsilon}{2}.$$

Since y is a limit point of $\mathcal{R}(\phi^t)$, there exists a chain recurrent point x such that $d(x,y) < \min\{\delta, \epsilon/2\}$. So, $(y, \phi^T(x); T)$ is an (ϵ, T)-chain from y to $\phi^T(x)$.

Since x is chain recurrent, there exists an $(\epsilon/2, 2T)$-chain

$$(x = x_0, ..., x_n = x; t_0, ..., t_{n-1})$$

from x to itself. Then

$$(\phi^T(x), x_1, ..., x_{n-1}; t_0 - T, t_1, ..., t_{n-2})$$

is an (ϵ, T)-chain from $\phi^T(x)$ to x_{n-1}.

By the triangle inequality,

$$d(\phi^{t_{n-1}}(x_{n-1}), y) < d(\phi^{t_{n-1}}(x_{n-1}), x) + d(x,y) < \frac{\epsilon}{2} + \frac{\epsilon}{2} = \epsilon.$$

Consequently, $(x_{n-1}, y; t_{n-1})$ is an (ϵ, T)-chain from x_{n-1} to y.

By concatenating these (ϵ, T)-chains,

$$(y, \phi^T(x), x_1, ..., x_{n-1}, y; T, t_0 - T, t_1, ..., t_{n-1})$$

is an (ϵ, T)-chain from y to itself. Therefore, $\mathcal{R}(\phi^t)$ is closed. □

Topological equivalences carry chain recurrent sets to chain recurrent sets under additional hypotheses on the topological equivalence.

Proposition 2.7.11. *Let (X, d_X) and (Y, d_Y) be metric spaces. If (h, ρ) is a topological equivalence from the flow $\phi^t : X \to X$ to the flow $\psi^t : Y \to Y$ such that $h : X \to Y$ is uniformly continuous,* $\sup\{\rho(T, z) \mid z \in X\}$ *exists for each $T > 0$, and* $\sup\{\tilde{\rho}(T, z) \mid z \in X\}$ *exists for each $T > 0$, then*

$$\mathcal{R}(\psi^t) = h\left(\mathcal{R}(\phi^t)\right).$$

Proof. We begin by showing that $h\left(\mathcal{R}(\phi^t)\right) \subseteq \mathcal{R}(\psi^t)$. Let $x \in \mathcal{R}(\phi^t)$. Let $\epsilon > 0$ and $T > 0$. We shall construct an (ϵ, T)-chain from $h(x)$ to itself.

Because h is uniformly continuous on X, there exists $\delta > 0$ such that if $d_X(z_1, z_2) < \delta$, then $d_Y(h(z_1), h(z_2)) < \epsilon$. Let $S = \sup\{\rho(T, z) \mid z \in X\}$. Since $x \in \mathcal{R}(\phi^t)$, there exists a (δ, S)-chain

$$(x = x_0, x_1, \ldots, x_n = x; s_0, \ldots, s_{n-1})$$

from x to itself. For each $i = 0, \ldots, n$ let $y_i = h(x_i)$. Since ρ is a reparametrization, there exists t_i so that $s_i = \rho(t_i, x_i)$ for each $= 0. \ldots, n-1$. We claim that

$$(h(x) = y_0, y_1, \ldots, y_n = h(x); t_0, \ldots, t_{n-1})$$

is an (ϵ, T)-chain from $h(x)$ to itself.

Since $s_i \geq S$ for $i = 0, \ldots, n-1$,

$$\rho(t_i, x_i) = s_i \geq S \geq \rho(T, x_i).$$

Because $\rho(\cdot, x_i)$ is an increasing function, $t_i \geq T$ for $i = 0, \ldots, n-1$. Since $d_X(\phi^{s_i}(x_i), x_{i+1}) < \delta$ for $i = 0, \ldots, n-1$,

$$\begin{aligned} d_Y(\psi^{t_i}(y_i), y_{i+1}) &= d_Y(\psi^{t_i}(h(x_i)), h(x_{i+1})) \\ &= d_Y(h(\phi^{\rho(t_i, x_i)}(x_i)), h(x_{i+1})) \\ &= d_Y(h(\phi^{s_i}(x_i)), h(x_{i+1})) \\ &< \epsilon \end{aligned}$$

for $i = 0, \ldots, n-1$. Thus,

$$(h(x) = y_0, y_1, \ldots, y_n = h(x); t_0, \ldots, t_{n-1})$$

is an (ϵ, T)-chain from $h(x)$ to itself so that $h(x) \in \mathcal{R}(\psi^t)$. Therefore, $h\left(\mathcal{R}(\phi^t)\right) \subseteq \mathcal{R}(\psi^t)$.

Applying the previous argument to the topological equivalence $(h^{-1}, \tilde{\rho})$ we obtain $h^{-1}(\mathcal{R}(\psi^t)) \subseteq \mathcal{R}(\phi^t)$. Thus, $\mathcal{R}(\psi^t) \subseteq h(\mathcal{R}(\phi^t))$. Therefore,

$$\mathcal{R}(\psi^t) = h\left(\mathcal{R}(\phi^t)\right). \qquad \square$$

As a corollary, the chain recurrent set satisfies the Topological Invariance Property for flows on compact spaces.

Corollary 2.7.12. *Let X and Y be metric spaces. Let $\phi^t : X \to X$ and $\psi^t : Y \to Y$ be flows. If X is compact and $h : X \to Y$ is a topological conjugacy from ϕ^t to ψ^t, then*

$$h(\mathcal{R}(\phi^t)) = \mathcal{R}(\psi^t).$$

Proof. Since X is compact, the homeomorphism h is uniformly continuous. Applying Proposition 2.7.11 with $\rho(t, x) = t$, we obtain $h(\mathcal{R}(\phi^t)) = \mathcal{R}(\psi^t)$. $\square$

The following example shows that compactness is necessary in the hypotheses of Corollary 2.7.12. If the phase space is not compact, then the chain recurrent set may depend on the metric even if two complete metrics induce the same topology.

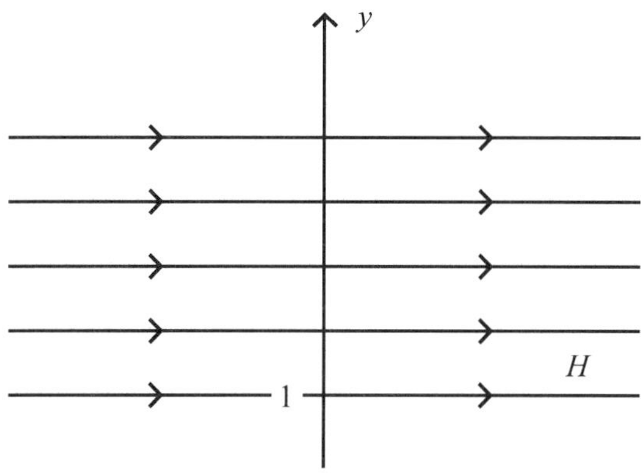

Figure 2.19. A phase portrait of the flow of Example 2.7.13

Example 2.7.13. Let $H = \{(x, y) \in \mathbb{R}^2 \mid y \geq 1\}$. Denote the metric on H induced by the standard metric on $\mathbb{R}^2$ by d_E, and denote the metric on H induced by the hyperbolic (Riemannian) metric on the upper half-plane $\mathbb{H}^2 = \{(x, y) \in \mathbb{R}^2 \mid y > 0\}$ by d_H. The metric d_H inherits the following properties from the hyperbolic metric:

$$d_H((x, y_0), (x, y_1)) = |y_1 - y_0|$$

and

$$d_H((x_0, y), (x_1, y)) = \frac{|x_1 - x_0|}{y}$$

for all (x, y_0), (x, y_1), (x_0, y), $(x_1, y) \in H$. Define $\phi : \mathbb{R} \times H \to H$ by

$$\phi(t, (x, y)) = (x + t, y).$$

Exercise 16 asks you to verify that ϕ is a flow, that the identity map $\iota : (H, d_E) \to (H, d_H)$ is a topological conjugacy from ϕ^t to itself, and that the chain recurrent set of ϕ^t with respect to d_E is the empty set. See Figure 2.19.

In contrast, we shall demonstrate that the chain recurrent set of ϕ^t with respect to d_H is H. Let $(x, y) \in H$. Let $\epsilon > 0$ and $T > 0$. There exists a positive integer m such that

$$m > \frac{2}{\epsilon}\left(\frac{2T}{\epsilon} - y\right).$$

Let $(x_0, y_0) = (x, y)$. Define

$$(x_{i+1}, y_{i+1}) = \begin{cases} (x_i + T, y_i + \epsilon/2) & \text{if } i = 0, \ldots, m-1, \\ (x_i - T, y_i) & \text{if } i = m, \ldots, 3m-1, \\ (x_i + T, y_i - \epsilon/2) & \text{if } i = 3m, \ldots, 4m-1. \end{cases}$$

Define $t_i = T$ for $i = 0, \ldots, 4m-1$. We claim that

$$\{(x_0, y_0), \ldots, (x_{4m}, y_{4m}); t_0, \ldots, t_{4m-1}\}$$

is an (ϵ, T)-chain from (x, y) to itself. If $i = 0, \ldots, m-1$, then

$$\begin{aligned} d_H(\phi^{t_i}(x_i, y_i), (x_{i+1}, y_{i+1})) &= d_H((x_i + T, y_i), (x_i + T, y_i + \epsilon/2)) \\ &= \epsilon/2 \\ &< \epsilon. \end{aligned}$$

If $i = m, \ldots, 3m-2$, then $y_i = y + m\epsilon/2 > 2T/\epsilon$. So,

$$\begin{aligned} d_H(\phi^{t_i}(x_i, y_i), (x_{i+1}, y_{i+1})) &= d_H((x_i + T, y_i), (x_i - T, y_i)) \\ &= \frac{2T}{y_i} \\ &< \epsilon. \end{aligned}$$

If $i = 3m-1, \ldots, 4m-1$, then

$$\begin{aligned} d_H(\phi^{t_i}(x_i, y_i), (x_{i+1}, y_{i+1})) &= d_H((x_i + T, y_i), (x_i + T, y_i - \epsilon/2)) \\ &= \epsilon/2 \\ &< \epsilon. \end{aligned}$$

Therefore, the chain recurrent set of ϕ^t with respect to the metric d_H is H.

The chain recurrent set contains the nonwandering set. Thus, Poincaré recurrent points and central points are chain recurrent.

Proposition 2.7.14. *The chain recurrent set of a flow on a metric space contains the nonwandering set of the flow.*

Proof. Let ϕ^t be a flow on a metric space (X, d). Let $x \in \Omega(\phi^t)$, and let $\epsilon > 0$ and $T > 0$. By the continuity of ϕ^T, there exists $\delta \in (0, \epsilon)$ such that if $d(x, y) < \delta$, then

$$d(\phi^T(x), \phi^T(y)) < \epsilon.$$

Denote the open δ-ball around x by $B(x, \delta)$. Since X is a metric space, X is Hausdorff. By Lemma 2.5.9 there is a real number $S \geq 2T$ with the property that

$$B(x, \delta) \cap \phi^S(B(x, \delta)) \neq \emptyset.$$

Consequently, there exists $y \in B(x, \delta)$ with $\phi^S(y) \in B(x, \delta)$. Since $d(x, y) < \delta$, the continuity of ϕ^T implies that

$$d(\phi^T(x), \phi^T(y)) < \epsilon.$$

Because $S \geq 2T$, we know that $S - T \geq T$. By the group property of flows,

$$d(\phi^{S-T}(\phi^T(y)), x) = d(\phi^S(y), x) < \delta < \epsilon.$$

Thus,

$$(x, \phi^T(y), x; T, S-T)$$

is an (ϵ, T)-chain from x to itself. Therefore, x is a chain recurrent point for ϕ^t, and

$$\Omega(\phi^t) \subseteq \mathcal{R}(\phi^t). \qquad \square$$

Let ϕ^t be a flow on a second-countable metric space. We have now established a chain of inclusions expressing more and more general modes of recurrence:

$$\text{Fix}(\phi^t) \subseteq \text{Per}(\phi^t) \subseteq \text{Rec}(\phi^t) \subseteq C(\phi^t) \subseteq \Omega(\phi^t) \subseteq \mathcal{R}(\phi^t).$$

Since the chain recurrent set contains the nonwandering set, the chain recurrent set contains the α- and ω-limit sets of each point in the phase space. Thus, the chain recurrent set satisfies the Accumulation Property.

Corollary 2.7.15. *If ϕ^t is a flow on a metric space X, then $\omega(x) \subseteq \mathcal{R}(\phi^t)$ and $\alpha(x) \subseteq \mathcal{R}(\phi^t)$ for all $x \in X$.*

Proof. By Propositions 2.5.14 and 2.7.14,

$$\alpha(x) \cup \omega(x) \subseteq \Omega(\phi^t) \subseteq \mathcal{R}(\phi^t)$$

for all $x \in X$. $\square$

2.7.4. Equivalent Formulations of Chain Equivalence. We now describe three alternative formulations of chain equivalence and prove that for a flow on a compact metric space these formulations coincide with Definition 2.7.3. For this purpose it is useful to provide definitions of ϵ-chain and chain equivalence with respect to a map.

Definition 2.7.16. Let (X, d) be a metric space, and let $f : X \to X$. Given $\epsilon > 0$ and $x, y \in X$, an **ϵ-chain** from x to y is a finite sequence

$$x = x_0, x_1, \ldots, x_{n-1}, x_n = y$$

in X such that

$$d(f(x_i), x_{i+1}) < \epsilon$$

for $i = 0, 1, 2, \ldots, n-1$. See Figure 2.20.

Definition 2.7.17. Let X be a metric space, and let $f : X \to X$. Two points $x, y \in X$ are called **chain equivalent** if for every $\epsilon > 0$ there exists an ϵ-chain from x to y and there exists an ϵ-chain from y to x.

We now describe three alternative formulations of chain equivalence which appear in the literature and prove that for a flow on a compact metric space these formulations coincide with Definition 2.7.3. The essence of this result appears in Hurley [**23**].

Theorem 2.7.18. *If ϕ^t is a flow on a compact metric space (X, d) and $x, y \in X$, then the following statements are equivalent.*

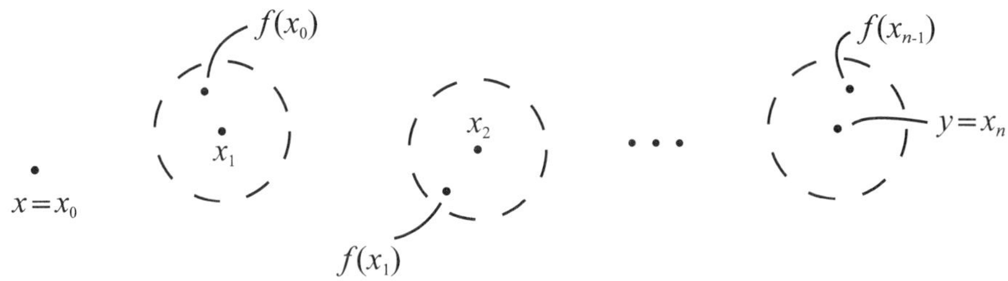

Figure 2.20. An ϵ-chain for a map

(i) *The points x and y are chain equivalent with respect to ϕ^t.*

(ii) *For every $\epsilon > 0$ and $T > 0$ there exists an $(\epsilon, 1)$-chain*

$$(x_0, \ldots, x_n; t_0, \ldots, t_{n-1})$$

from x to y such that

$$t_0 + \cdots + t_{n-1} \geq T,$$

and there exists an $(\epsilon, 1)$-chain

$$(y_0, \ldots, y_n; s_0, \ldots, s_{m-1})$$

from y to x such that

$$s_0 + \cdots + s_{m-1} \geq T.$$

(iii) *For every $\epsilon > 0$ there exists an $(\epsilon, 1)$-chain from x to y and an $(\epsilon, 1)$-chain from y to x.*

(iv) *The points x and y are chain equivalent with respect to ϕ^1.*

In the spirit of Robinson [**37**] p. 151, Theorem 2.7.18 (ii) asserts that two points x and y are chain equivalent exactly when, given any specified time T, there are $(\epsilon, 1)$-chains from x to y and from y to x so that the total time along each chain is at least T. Following Franks [**14**] p. 1, Theorem 2.7.18 (iii) removes the condition from part (ii) concerning the total time along each chain. Theorem 2.7.18 (iv) states that two points are chain equivalent with respect to a flow exactly when they are chain equivalent with respect to the time-one map of the flow.

In preparation for the proof of Theorem 2.7.18 we establish two lemmas. The first shows that if there exists an $(\epsilon, 1)$-chain from x to y and an $(\epsilon, 1)$-chain from y to x, then there is an $(\epsilon, 1)$-chain from x to y in which all of the times are at most two and the total time along the chain is nearly an integer. The proof exploits Proposition B.0.9 regarding irrational rotations of the circle.

Lemma 2.7.19. *Let ϕ^t be a flow on a metric space (X, d). If $x, y \in X$ and for every $\epsilon > 0$ there exists an $(\epsilon, 1)$-chain from x to y and there exists an $(\epsilon, 1)$-chain from y to x, then for every $\epsilon > 0$ and $\delta > 0$ there exists a positive integer K and an $(\epsilon, 1)$-chain*

$$(x = x_0, x_1, \dots, x_n = y; t_0, t_1, \dots, t_{n-1})$$

from x to y such that

$$1 \leq t_i < 2$$

for $i = 0, 1, 2, \dots, n-1$, and if $T = t_0 + t_1 + \cdots + t_{n-1}$, then

$$|T - K| < \delta.$$

Proof. Let $\epsilon > 0$ and $\delta > 0$. By hypothesis there is an $(\epsilon, 1)$-chain

$$(x = w_0, w_1, \dots, w_n = y; r_0, r_1, \dots, r_{n-1})$$

from x to y. Without loss of generality, we may assume that $r_i \in [1, 2)$ for $i = 0, 1, 2, \dots, n-1$. If not, then insert points and times into this chain, such as $\phi^1(x)$ and time 1, and reindex the points and times in the chain. Let $R = r_0 + r_1 + \cdots + r_{n-1}$. Similarly, there is an $(\epsilon, 1)$-chain

$$(y = z_0, z_1, \dots, z_m = x; s_0, s_1, \dots, s_{m-1})$$

such that $s_i \in [1, 2)$ for $i = 0, 1, 2, \dots m-1$. Let $S = s_0 + s_1 + \cdots + s_{m-1}$. By the continuity of ϕ^t there exists $\eta > 0$ such that if $|t - r_0| < \eta$, then

$$d(\phi^t(w_0), \phi^{r_0}(w_0)) < \epsilon - d(\phi^{r_0}(w_0), w_1).$$

By the density of the irrational numbers among the real numbers there exists an $\bar{r}_0 \in [r_0, r_0 + \eta)$ such that $R + S - r_0 + \bar{r}_0$ is irrational. By the triangle inequality,

$$\begin{aligned} d(\phi^{\bar{r}_0}(w_0), w_1) &\leq d(\phi^{\bar{r}_0}(w_0), \phi^{r_0}(w_0)) + d(\phi^{r_0}(w_0), w_1) \\ &< \epsilon - d(\phi^{r_0}(w_0), w_1) + d(\phi^{r_0}(w_0), w_1) \\ &= \epsilon. \end{aligned}$$

Thus, $(x = w_0, w_1, \dots, w_n = y; \bar{r}_0, r_1, \dots, r_{n-1})$ is an $(\epsilon, 1)$-chain from x to y.

Let $\bar{R} = \bar{r}_0 + r_1 + \cdots + r_{n-1}$. Since $\bar{R} + S$ is irrational, Proposition B.0.9 implies that there exist positive integers J and K so that

$$|\bar{R} + J(\bar{R} + S) - K| < \delta,$$

We shall construct an $(\epsilon, 1)$-chain from x to y so that the sum of times in the chain is $\bar{R} + J(\bar{R} + S)$. Begin with the chain

$$(x = w_0, w_1, \dots, w_n = y; \bar{r}_0, r_1, \dots, r_{n-1}),$$

and alternately concatenate the chains

$$(y = z_0, z_1, \dots, z_m = x; s_0, s_1, \dots, s_{m-1})$$

and

$$(x = w_0, w_1, \dots, w_n = y; \bar{r}_0, r_1, \dots, r_{n-1})$$

a total of J times to obtain an $(\epsilon, 1)$-chain

$$(w_0, \dots, w_n = \underbrace{z_0, \dots, z_m, w_1, \dots, y}_{J \text{ times}}; \bar{r}_0, \dots, r_{n-1}, \underbrace{s_0, \dots, s_{m-1}, \bar{r}_0, \dots, r_{n-1}}_{J \text{ times}})$$

from x to y. The sum $T = \bar{R} + J(\bar{R} + S)$ of the times in this chain has the property that

$$|T - K| < \delta.$$

□

Often we use the fact that the flow ϕ^t is continuous on $\mathbb{R} \times X$ together with the compactness of X to obtain local uniform continuity. The next technical lemma is such a result.

Lemma 2.7.20. *Let ϕ^t be a flow on a compact metric space (X, d). Then for every $\epsilon > 0$ there is a $\delta > 0$ so that if $|t| < \delta$, then*

$$d(\phi^t(x), x) < \epsilon$$

for every $x \in X$.

Proof. Let $\epsilon > 0$. The compactness of X implies that $[-1, 1] \times X$ is compact. So, since ϕ^t is continuous, ϕ^t is uniformly continuous on $[-1, 1] \times X$. Thus, there exists $\delta \in (0, 1)$ such that if $|t - s| < \delta$ and $d(x, z) < \delta$, then

$$d\left(\phi^t(x), \phi^s(z)\right) < \epsilon.$$

In particular, $d(x, x) = 0 < \delta$ for each $x \in X$. Therefore, if $t \in (-\delta, \delta)$, then $x \in X$. □

We will now prove Theorem 2.7.18

Proof. We shall prove that

$$\text{(i)} \Rightarrow \text{(ii)} \Rightarrow \text{(iii)} \Rightarrow \text{(iv)} \Rightarrow \text{(i)}.$$

(1) The proof that (i) implies (ii) is immediate.

(2) The proof that (ii) implies (iii) is also immediate.

(3) To prove that (iii) implies (iv), assume that for each $\epsilon > 0$ there exists an $(\epsilon, 1)$-chain from x to y and there exists an $(\epsilon, 1)$-chain from y to x. We must prove that for each $\epsilon > 0$ there exists an ϵ-chain from x to y with respect to ϕ^1. The argument that there exists an ϵ-chain from y to x with respect to ϕ^1 is similar.

Let $\epsilon > 0$. By Lemma 2.7.20 there exists $\delta_1 \in (0, 1)$ such that if $t \in (-\delta_1, \delta_1)$, then

$$d(\phi^t(z), z) < \frac{\epsilon}{4} \tag{2.1}$$

for all $z \in X$. By the continuity of ϕ^t and the compactness of $[0, 3] \times X$, there exists $\delta_2 > 0$ such that if $z_1, z_2 \in X$ and $d(z_1, z_2) < \delta_2$, then

$$d(\phi^t(z_1), \phi^t(z_2)) < \frac{\epsilon}{2} \tag{2.2}$$

for all $t \in [0, 3]$.

Let $\eta = \min\{\epsilon/4, \delta_2\}$. By Lemma 2.7.19 there exists an $(\eta, 1)$-chain

$$(x = x_0, x_1, \ldots, x_n = y; t_0, t_1, \ldots, t_{n-1})$$

from x to y and there exists a positive integer K such that

$$1 \leq t_i < 2$$

for $i = 0, 1, 2, \ldots, n-1$, and

$$|T - K| < \delta_1$$

where $T = t_0 + t_1 + \ldots + t_{n-1}$. Denote the chain

$$(x = x_0, x_1, \ldots, x_n = y; t_0, t_1, \ldots, t_{n-1})$$

by $\mathcal{C}$. From $\mathcal{C}$ we shall obtain an ϵ-chain

$$x = y_0, y_1, \ldots, y_K = y$$

with respect to ϕ^1.

Let $y_0 = x_0 = x$. The new chain coincides with $\mathcal{C}$ for as many steps as $t_i = 1$. The modifications to $\mathcal{C}$ begin at the first step at which $t_i > 1$. Without loss of generality, we may assume that $t_0 > 1$.

Let $y_1 = \phi^1(y_0)$. Accordingly, $d(\phi^1(y_0), y_1) = 0 < \epsilon$.

The strategy for defining $y_2, \ldots, y_K$ is to account for the time elapsed along the new chain versus the time remaining along the chain $\mathcal{C}$. On the arc from y_0 to $\phi^1(y_0)$, we have used one unit of time. The total time remaining along $\mathcal{C}$ is $T - 1$, while the time remaining along the arc from x_0 to $\phi^{t_0}(x_0)$ is $t_0 - 1$. Since $1 < t_0 < 2$, we have $0 < t_0 - 1 < 1$. To define y_2, jump to the orbit of x_1 and flow for $1 - (t_0 - 1) = 2 - t_0$ units of time. Observe that $0 < 2 - t_0 < 1 \leq t_1$. It follows from the choice of η that

$$d(\phi^{t_0}(x_0), x_1) < \delta_2.$$

Consequently, inequality (2.2) implies that

$$d(\phi^{2-t_0}(\phi^{t_0}(x_0)), \phi^{2-t_0}(x_1)) < \frac{\epsilon}{2}.$$

Let $y_2 = \phi^{2-t_0}(x_1)$. From the definitions of y_0, y_1 and y_2 we obtain

$$\begin{aligned} d(\phi^1(y_1), y_2) &= d(\phi^2(y_0), \phi^{2-t_0}(x_1)) \\ &= d(\phi^{2-t_0}(\phi^{t_0}(x_0)), \phi^{2-t_0}(x_1)) \\ &< \frac{\epsilon}{2}. \end{aligned}$$

A total of 2 units of time have elapsed along the chain $\mathcal{C}$, and $y_2 = \phi^{2-t_0}(x_1)$ lies on the arc joining x_1 to $\phi^{t_1}(x_1)$. The time remaining on this arc is $t_1 - (2 - t_0) = t_0 + t_1 - 2$.

To define y_3 there are three cases to consider. If there is more than one unit of time remaining on the arc joining x_1 to $\phi^{t_1}(x_1)$, then continue to follow the orbit of x_1 for time one. If there is exactly one unit of time remaining, then jump to the arc of the orbit joining x_2 and $\phi^{t_2}(x_2)$. If there is less than one unit of time remaining, then to recover an additional $1-(t_0+t_1-2) = 3-(t_0+t_1)$ units of time, jump to an appropriate point on the orbit of x_2.

More precisely, the three cases are:

(a) If $t_0 + t_1 - 2 > 1$, then let $y_3 = \phi^1(y_2)$, in which case

$$d(\phi^1(y_2), y_3) = 0 < \epsilon.$$

(b) If $t_0 + t_1 - 2 = 1$, that is, $t_0 + t_1 = 3$, then let $y_3 = x_2$. Since $y_2 = \phi^{2-t_0}(x_1)$, the properties of the chain $\mathcal{C}$ guarantee that

$$d(\phi^1(y_2), y_3) = d(\phi^{t_1}(x_1), x_2) < \epsilon.$$

(c) If $t_0 + t_1 - 2 < 1$, then define $y_3 = \phi^{3-(t_0+t_1)}(x_2)$. Since

$$d(\phi^{t_1}(x_1), x_2) < \delta_2,$$

and $3 - (t_0 + t_1) < 1$, we conclude from inequality (2.2) that

$$\begin{aligned} d(\phi^1(y_2), y_3) &= d(\phi^{3-t_0}(x_1), \phi^{3-(t_0+t_1)}(x_2)) \\ &= d(\phi^{3-(t_0+t_1)}(\phi^{t_1}(x_1)), \phi^{3-(t_0+t_1)}(x_2)) \\ &< \frac{\epsilon}{2}. \end{aligned}$$

The preceding argument suggests how to define the rest of the points in the chain $y_0, \ldots, y_K$. Suppose that we have defined $y_0, \ldots, y_j$. Then y_j lies on the arc of the orbit of x_i joining x_i with $\phi^{t_i}(x_i)$ for some $i = 0, 1, 2, \ldots, n-1$. If there is more than one unit of time remaining on the arc starting at x_i, then define y_{j+1} to be on the same arc at time one away from y_j. If there is exactly one unit of time left on the arc starting at x_i, then define y_{j+1} to be x_{i+1}. Finally, if there is less than one unit of time remaining on the arc starting at x_i, then define y_{j+1} to be an appropriate point on the orbit of x_{i+1}.

More precisely, suppose we have defined y_j. There exists a natural number i such that $0 \leq i \leq n-1$ and there exists $\tau \in [0, t_i)$ such that $y_j = \phi^\tau(x_i)$.

(a) If $t_i - \tau > 1$, then define $y_{j+1} = \phi^1(y_j)$.
(b) If $t_i - \tau = 1$, then define $y_{j+1} = x_{i+1}$.
(c) If $t_i - \tau < 1$, then define $y_{j+1} = \phi^{1-(t_i - \tau)}(x_{i+1})$.

The details verifying that

$$d(\phi^1(y_j), y_{j+1}) < \epsilon$$

in each of these cases are similar to those of our earlier argument.

Continue in this manner until we have defined y_{K-1}. The time elapsed along the new ϵ-chain is $K - 1$. Recall that the total time along the chain $\mathcal{C}$ is T. Thus, the time remaining is

$$T - (K - 1) = 1 + (T - K)$$

where $|T - K| < \delta_1$.

Let $y_K = y$. We shall prove that $d(\phi^1(y_{K-1}), y) < \epsilon$. The amount of time remaining along the chain $\mathcal{C}$ is within δ_1 of 1 and $0 < \delta_1 < 1$. So, y_{K-1} lies either on the arc of the orbit of x_{n-1} joining x_{n-1} with $\phi^{t_{n-1}}(x_{n-1})$ or lies on the arc of the orbit of x_{n-2} joining x_{n-2} with $\phi^{t_{n-2}}(x_{n-2})$.

If y_{K-1} lies on the arc of the orbit of x_{n-1} joining x_{n-1} to $\phi^{t_{n-1}}(x_{n-1})$, then there exists $\tau \in [0, t_{n-1})$ such that $y_{K-1} = \phi^\tau(x_{n-1})$, and the remaining time along this arc is $t_{n-1} - \tau = 1 + (T - K)$. Therefore,

$$\begin{aligned} d(\phi^1(y_{K-1}), y) &= d(\phi^{\tau+1}(x_{n-1}), y) \\ &= d(\phi^{t_{n-1}-(T-K)}(x_{n-1}), y) \\ &\leq d(\phi^{t_{n-1}-(T-K)}(x_{n-1}), \phi^{t_{n-1}}(x_{n-1})) \\ &\qquad + d(\phi^{t_{n-1}}(x_{n-1}), y). \end{aligned}$$

By inequality (2.1) and the conditions on the chain $\mathcal{C}$, we obtain

$$d(\phi^1(y_{K-1}), y) < \frac{\epsilon}{4} + \frac{\epsilon}{2} < \epsilon.$$

If y_{K-1} lies on the arc of the orbit of x_{n-2} joining x_{n-2} to $\phi^{t_{n-2}}(x_{n-2})$, then there exists $\tau \in [0, t_{n-2})$ such that $y_{K-1} = \phi^\tau(x_{n-2})$. The total remaining time, on the one hand, is the time remaining on the arc joining x_{n-2} with $\phi^{t_{n-2}}(x_{n-2})$ plus the time along the arc joining x_{n-1} with $\phi^{t_{n-1}}(x_{n-1})$, and, on the other hand, is the total time along the chain minus the time elapsed so far. So,

$$t_{n-2} - \tau + t_{n-1} = T - (K - 1).$$

Consequently,

$$\tau + 1 - t_{n-2} = t_{n-1} - (T - K).$$

Since $1 \le t_{n-1} < 2$ and $|T - K| < \delta_1 < 1$, we obtain

$$\tau + 1 - t_{n-2} = t_{n-1} - (T - K) \le 3.$$

By inequality (2.2), $d(\phi^{t_{n-2}}(x_{n-2}), x_{n-1}) < \eta \le \delta_2$ implies that

$$d(\phi^{\tau+1}(x_{n-2}), \phi^{t_{n-1}-(T-K)}(x_{n-1})) < \frac{\epsilon}{2}.$$

By inequality (2.1), $|T - K| < \delta_1$ implies that

$$d(\phi^{t_{n-1}-(T-K)}(x_{n-1}), \phi^{t_{n-1}}(x_{n-1})) < \frac{\epsilon}{4}.$$

Because $\mathcal{C}$ is an $(\eta, 1)$-chain,

$$d(\phi^{t_{n-1}}(x_{n-1}), y) < \eta \le \frac{\epsilon}{4}.$$

Thus,

$$\begin{aligned} d(\phi^1(y_{K-1}), y) &= d(\phi^{\tau+1}(x_{n-2}), y) \\ &\le d(\phi^{\tau+1}(x_{n-2}), \phi^{t_{n-1}-(T-K)}(x_{n-1})) \\ &\qquad + d(\phi^{t_{n-1}-(T-K)}(x_{n-1}), \phi^{t_{n-1}}(x_{n-1})) \\ &\qquad + d(\phi^{t_{n-1}}(x_{n-1}), y) \\ &< \frac{\epsilon}{2} + \frac{\epsilon}{4} + \frac{\epsilon}{4} = \epsilon. \end{aligned}$$

Therefore, the sequence $x = y_0, y_1, \ldots, y_{K-1}, y_K = y$ satisfies

$$d(\phi^1(y_i), y_{i+1}) < \epsilon$$

for $i = 0, 1, 2, \ldots, K - 1$. Thus, $x = y_0, y_1, \ldots, y_{K-1}, y_K = y$ is an ϵ-chain from x to y with respect to ϕ^1. Similarly, there exists an ϵ-chain from y to x. Therefore, x and y are chain equivalent with respect to ϕ^1.

(4) To prove that (iv) implies (i), assume that x and y are chain equivalent with respect to ϕ^1. Let $\epsilon > 0$ and $T > 0$. Without loss of generality, we may assume that T is an integer.

Let $\delta_0 = \epsilon$. Since X is compact and ϕ^1 is continuous, the function ϕ^1 is uniformly continuous on X. Thus, there exists $\delta_1 \in (0, \epsilon)$ such that if $w_1, w_2 \in X$ and $d(w_1, w_2) < \delta_1$, then

$$d(\phi^1(w_1), \phi^1(w_2)) < \frac{\delta_0}{2}.$$

Similarly, for each $j = 2, 3, 4, \ldots, 2T - 1$ there exists $\delta_j \in (0, \epsilon)$ such that if $w_1, w_2 \in X$ and $d(w_1, w_2) < \delta_j$, then

$$d(\phi^1(w_1), \phi^1(w_2)) < \frac{\delta_{j-1}}{2}.$$

Define

$$\eta = \min\left\{\frac{\delta_0}{2}, \dots, \frac{\delta_{2T-1}}{2}\right\}.$$

By hypothesis there exist η-chains

$$x = x_0, x_1, \dots, x_{K-1}, x_K = y$$

and

$$y = y_0, y_1, \dots, y_{L-1}, y_L = x.$$

Begin with the sequence $x_0, x_1, \dots, x_{K-1}$ and concatenate T copies of $y_0, y_1, \dots, y_{L-1}, y_L = x_0, x_1, \dots, x_{K-1}$. End the sequence with $x_K = y$ to obtain a sequence

$$x = x_0, x_1, \dots, x_K = \underbrace{y = y_0, \dots, y_L = x = x_0, x_1, \dots, x_{K-1}}_{T \text{ times}}, y$$

in X with $K + T(K+L) + 1$ entries.

By labeling this new sequence we have an η-chain

$$x = w_0, w_1, \dots, w_{K+T(K+L)} = y$$

with respect to ϕ^1.

Since K, L and T are positive integers, there exist positive integers n and r such that

$$K + T(K+L) = nT + r$$

and $T \le r < 2T$.

Define

$$z_i = \begin{cases} w_{iT} & \text{if } 0 \le i < n, \\ y & \text{if } i = n \end{cases}$$

and

$$t_i = \begin{cases} T & \text{if } 0 \le i < n-1, \\ r & \text{if } i = n-1 \end{cases}$$

for $i = 0, 1, 2, \dots, n$.

We claim that $(z_0, \dots, z_n; t_0, \dots, t_{n-1})$ is an (ϵ, T)-chain from x to y. Let $i = 0, 1, 2, \dots, n$. Observe that by the choice of η,

$$d(\phi^1(w_{iT}), w_{iT+1}) < \eta \le \delta_{r-1}.$$

Assume that for some integer k such that $2 \le k \le r$ we have

$$d(\phi^{k-1}(w_{iT}), w_{iT+k-1}) < \eta \le \delta_{r-k+1}.$$

(We have just shown that this is true for $k = 2$.) From the choice of δ_{r-k+1} we find that

$$d(\phi^k(w_{iT}), \phi^1(w_{iT+k-1})) < \frac{\delta_{r-k}}{2}.$$

By the triangle inequality,

$$\begin{aligned} d(\phi^k(w_{iT}), w_{iT+k}) &\leq d(\phi^k(w_{iT}), \phi^1(w_{iT+k-1})) \\ &\quad + d(\phi^1(w_{iT+k-1}), w_{iT+k}) \\ &< \frac{\delta_{r-k}}{2} + \eta \\ &\leq \frac{\delta_{r-k}}{2} + \frac{\delta_{r-k}}{2} \\ &= \delta_{r-k}. \end{aligned}$$

Thus, if $k = T$, then

$$d(\phi^T(z_i), z_{i+1}) = d(\phi^T(w_{iT}), w_{(i+1)T}) < \delta_{r-T} \leq \epsilon$$

for $i = 0, 1, 2, \ldots, n-2$. If $k = r$, then

$$d(\phi^r(z_{n-1}), z_n) = d(\phi^r(w_{nT}), w_{nT+r}) < \delta_0 = \epsilon.$$

Hence, $(z_0, \ldots, z_n, t_0, \ldots, t_{n-1})$ is an (ϵ, T)-chain from x to y.

The construction of an (ϵ, T)-chain from y to x is similar. Therefore, (iv) implies (i).

We have now shown that statements (i)–(iv) are equivalent. □

2.7.5. The Restriction Property of the Chain Recurrent Set. Now we show that if ϕ^t is a flow on a compact metric space, then

$$\mathcal{R}(\phi^t|_{\mathcal{R}(\phi^t)}) = \mathcal{R}(\phi^t),$$

so that the chain recurrent set satisfies the Restriction Property. In other words, for each $x \in \mathcal{R}(\phi^t)$, $\epsilon > 0$ and $T > 0$, there exists an (ϵ, T)-chain

$$(x = x_0, x_1, \ldots, x_{n-1}, x_n = x; t_0, \ldots, t_{n-1})$$

from x to itself such that $x_i \in \mathcal{R}(\phi^t)$ for $i = 0, 1, 2, \ldots, n$. The idea of the proof is due to Robinson [**36**], although we employ the Hausdorff metric to streamline its presentation.

Proposition 2.7.21. *If ϕ^t is a flow on a compact metric space (X, d), then*

$$\mathcal{R}(\phi^t|_{\mathcal{R}(\phi^t)}) = \mathcal{R}(\phi^t).$$

Proof. The inclusion $\mathcal{R}(\phi^t|_{\mathcal{R}(\phi^t)}) \subseteq \mathcal{R}(\phi^t)$ is immediate. Thus, it suffices to prove that

$$\mathcal{R}(\phi^t) \subseteq \mathcal{R}(\phi^t|_{\mathcal{R}(\phi^t)}).$$

Let $x \in \mathcal{R}(\phi^t)$. By Theorem 2.7.18 (iv), for each positive integer n there exists a sequence

$$x = x_{n,0}, x_{n,1}, x_{n,2}, \ldots, x_{n,k_n} = x$$

such that

$$d(\phi^1(x_{n,i}), x_{n,i+1}) < \frac{1}{n}$$

for $i = 0, 1, 2, \ldots, k_n - 1$. Define

$$C_n = \{x_{n,0}, \ldots, x_{n,k_n}\}.$$

For each positive integer n, the set C_n is a compact subset of X.

Let $\mathcal{H}$ be the collection of nonempty compact subsets of X. Since X is compact, $\mathcal{H}$ coincides with the collection of nonempty closed bounded subsets of X. Define $D : \mathcal{H} \times \mathcal{H} \to \mathbb{R}$ by

$$D(A, B) = \sup\{d(a, B), d(b, A) \mid a \in A,\ b \in B\}.$$

By Proposition C.0.10 the function D is a metric on $\mathcal{H}$. We call this metric the Hausdorff metric. By Proposition C.0.15, the space $\mathcal{H}$ is compact with respect to D. Thus, the sequence $\{C_n\}$ in $\mathcal{H}$ contains a subsequence $\{C_{n_j}\}$ which converges to some $C \in \mathcal{H}$. By reindexing this subsequence, we may assume, without loss of generality, that $\{C_n\}$ converges to C. For each $p \in C$ and $\epsilon > 0$ we shall obtain a sequence

$$p = p_0, p_1, \ldots, p_{m-1}, p_m = p$$

in C such that

$$d(\phi^1(p_i), p_{i+1}) < \epsilon$$

for $i = 0, 1, 2, \ldots, m - 1$. so that every point in C is chain recurrent.

Let $p \in C$ and $\epsilon > 0$. The compactness of X and continuity of ϕ^1 imply that ϕ^1 is uniformly continuous. Thus, there exists $\delta \in (0, \epsilon/3)$ such that if $y_1, y_2 \in X$ and $d(y_1, y_2) < \delta$, then

$$d(\phi^1(y_1), \phi^1(y_2)) < \frac{\epsilon}{3}.$$

Since $\{C_n\}$ converges to C there exists a positive integer n such that

$$\frac{1}{n} < \frac{\epsilon}{3},$$

and

$$D(C_n, C) < \delta.$$

So, there exists $x_{n,j} \in C_n$ such that

$$d(x_{n,j}, p) < \delta.$$

Cyclically permute the elements of C_n by defining

$$y_i = \begin{cases} x_{n,i+j} & \text{if } i = 0, 1, 2, \ldots, k_n - j - 1, \\ x_{n,i+j-k_n} & \text{if } i = k_n - j, \ldots, k_n. \end{cases}$$

Since $D(C_n, C) < \delta$, there exists $p_j \in C$ such that

$$d(y_j, p_j) < \delta$$

for each $j = 1, \ldots, k_n - 1$. Let

$$p_0 = p_{k_n} = p.$$

We will show that $p = p_0, p_1, \ldots, p_{k_n} = p$ is an ϵ-chain in C with respect to ϕ^1.

Let $j \in \{0, 1, 2, \ldots, k_n - 1\}$. By the uniform continuity of ϕ^1 we obtain

$$d(\phi^1(p_j), \phi^1(y_j)) < \frac{\epsilon}{3}.$$

From the definition of $y_1, \ldots, y_{k_n}$,

$$d(\phi^1(y_j), y_{j+1}) < \delta < \frac{\epsilon}{3}.$$

The choice of p_{j+1} guarantees that

$$d(y_{j+1}, p_{j+1}) < \delta < \frac{\epsilon}{3}.$$

The triangle inequality implies that

$$\begin{aligned} d(\phi^1(p_j), p_{j+1}) &\leq d(\phi^1(p_j), \phi^1(y_j)) + d(\phi^1(y_j), y_{j+1}) + d(y_{j+1}, p_{j+1}) \\ &< \frac{\epsilon}{3} + \frac{\epsilon}{3} + \frac{\epsilon}{3} \\ &= \epsilon. \end{aligned}$$

Consequently, $p_0, p_1, p_2, \ldots, p_{k_n}$ is an ϵ-chain in C with respect to ϕ^1 from p to itself. Since p is an arbitrary element of C, Theorem 2.7.18 (iv) implies that

$$C \subseteq \mathcal{R}(\phi^t).$$

Since $x \in C_n$ for each positive integer n, we have $x \in C$. Thus, for each $\epsilon > 0$ and $T > 0$ there exists an (ϵ, T)-chain in $\mathcal{R}(\phi^t)$ from x to itself. Hence,

$$x \in \mathcal{R}(\phi^t|_{\mathcal{R}(\phi^t)}).$$

Therefore,

$$\mathcal{R}(\phi^t|_{\mathcal{R}(\phi^t)}) = \mathcal{R}(\phi^t).$$ □

Example 2.7.22. Let ψ^t be the arrested logistic rotation-dilation on the closed unit disk $\mathbb{D}$ in the complex plane. Example 2.7.9 shows that

$$\mathcal{R}(\psi^t) = \partial\mathbb{D} \cup \{0\}.$$

By restricting ψ^t to its chain recurrent set, we obtain the arrested rotation on S^1 described in Example 2.7.6 together with a fixed point at $\{0\}$. Therefore,

$$\mathcal{R}(\psi^t|_{\mathcal{R}(\psi^t)}) = \partial\mathbb{D} \cup \{0\} = \mathcal{R}(\psi^t).$$

In contrast,

$$\Omega(\psi^t|_{\Omega(\psi^t)}) = \{-1, 0, 1\} \neq \Omega(\psi^t)$$

by Example 2.5.16.

2.8. A Summary of Recurrent Sets and Their Properties

Let ϕ^t be a flow on a second-countable compact metric space. In the table below, each box corresponds to a type of recurrence and to a desirable property of a recurrent set. The symbol $\star$ in a box indicates that the type of recurrence satisfies the corresponding property. The symbol – in a box indicates that the type of recurrence does not satisfy the corresponding property. Each box contains a reference to the proposition, corollary or example that proves or disproves the corresponding property. Chapter 3 addresses the Decomposition Property.

	$\mathrm{Fix}(\phi^t)$	$\mathrm{Per}(\phi^t)$	$\mathrm{Rec}(\phi^t)$	$C(\phi^t)$	$\Omega(\phi^t)$	$\mathcal{R}(\phi^t)$
Flow Invariance	$\star$ 2.1.2	$\star$ 2.2.12	$\star$ 2.4.5	$\star$ 2.6.6	$\star$ 2.5.5	$\star$ 3.3.8
Closure	$\star$ 2.1.7	– 2.2.14	– 2.4.6	$\star$ 2.6.6	$\star$ 2.5.6	$\star$ 2.7.10
Top. Invariance	$\star$ 2.1.8	$\star$ 2.2.15	$\star$ 2.4.7	$\star$ 2.6.9	$\star$ 2.5.10	$\star$ 2.7.12
Accumulation	– 2.3.8	– 2.3.9	– 2.4.8	– 2.6.10	$\star$ 2.5.14	$\star$ 2.7.15
Restriction	$\star$ 2.1.9	$\star$ 2.2.17	$\star$ 2.4.9	$\star$ 2.6.5	– 2.5.16	$\star$ 2.7.21

2.9. Exercises

(1) Find the set of fixed points of the flow corresponding to the Lorenz system of differential equations

$$\begin{aligned} \dot{x} &= -10x + 10y, \\ \dot{y} &= 28x - y - xz, \\ \dot{z} &= \frac{8}{3}z + xy \end{aligned}$$

on $\mathbb{R}^3$.

(2) Let p be a periodic point with period T for a flow ϕ^t such that $\{p\}$ is closed.

(a) Prove that $T = \inf\{|\tau - \sigma| \mid \tau \neq \sigma \text{ and } \phi^\tau(p) = \phi^\sigma(p) = p\}$.

(b) Prove that $T = 0$ or $T \in \{\tau > 0 \mid \phi^\tau(p) = p\}$.

(3) Prove that every flow on S^1 has a periodic point.

(4) Consider the flow corresponding to the system of differential equations

$$\begin{aligned} \dot{x}_1 &= c_1, \\ \dot{x}_2 &= c_2 \end{aligned}$$

on the torus $\mathbb{T}^2$. Prove that if $c_1 = 0$, then every point is periodic with period $1/|c_2|$, and if $c_2 = 0$ and $c_1 \neq 0$, then every point is periodic with period $1/|c_1|$.

(5) An **action** of a group G on a set S is a function $G \times S \to S$, denoted $(g, x) \mapsto gx$, such that for every $x \in S$ and $g_1, g_2 \in G$: $ex = x$ and $(g_1g_2)x = g_1(g_2x)$ where e is the identity element of G. The **stabilizer** of $x \in S$ is the set $\text{stab}(x) = \{g \in G \,|\, gx = x\}$. Chapter 1, Exercise 5 shows that a flow ϕ^t on a topological space X is a continuous action of the additive topological group $\mathbb{R}$ on X.
  (a) Prove that $\text{Per}(\phi^t) = \{x \in X \mid \text{stab}(x) \neq \{0\}\}$.
  (b) Let $x \in \text{Per}(\phi^t)$. Describe $\text{stab}(x)$.

(6) Prove or disprove: Every periodic orbit of a flow is a singleton fixed point or is homeomorphic to S^1.

(7) Let ϕ^t be the flow of the system of differential equations

$$\begin{aligned} \dot{x} &= \frac{4xy}{(x^2+y^2+1)^2}, \\ \dot{y} &= \frac{2(y^2-x^2+1)}{(x^2+y^2+1)^2} \end{aligned}$$

on $\mathbb{R}^2$. Show that the periodic set of ϕ^t is not closed. *Hint:* The orbits of ϕ^t are the level sets of $F : \mathbb{R}^2 \to \mathbb{R}$ defined by

$$F(x, y) = \frac{2x}{x^2+y^2+1}.$$

(8) Provide an example of a flow ϕ^t on the 2-sphere $S^2 = \{(x, y, z) \in \mathbb{R}^3 \mid x^2 + y^2 + z^2 = 1\}$ such that the set of periodic points of ϕ^t is not closed. *Hint:* Reparametrize the flow in Exercise 7 so that it extends to the one-point compactification of $\mathbb{R}^2$. Map the extended flow to S^2 via the inverse of stereographic projection.

(9) Let $\phi^t : [0, 1] \to [0, 1]$ be the flow associated to the logistic differential equation

$$\dot{x} = x(1 - x)$$

on $[0, 1]$. Find $\alpha(p)$ and $\omega(p)$ for all $p \in [0, 1]$ with respect to ϕ^t using the explicit expression

$$\phi^t(x) = \frac{xe^t}{1 + x(e^t - 1)}.$$

Compare with Example 2.3.4.

(10) Consider the flow ϕ^t corresponding to the system of differential equations

$$\begin{aligned} \dot{r} &= \frac{r}{1+r}, \\ \dot{\theta} &= \frac{1}{1+r} \end{aligned}$$

in polar coordinates on $\mathbb{R}^2$. Let $I = (-1, 1)$.

(a) Prove that if $p \in \mathbb{R}^2 \setminus \{(0,0)\}$, then the ω-limit set of p with respect to ϕ^t is empty. Compare with Proposition 2.3.6 (vi).

(b) Let $Z = \{(x,y) \in \mathbb{R}^2 \mid x \in I\}$. Define $g : Z \to \mathbb{R}^2$ by

$$g(x,y) = \left(\frac{x}{1-x^2}, y \right).$$

Prove that g is invertible.

(c) Define $\psi : \mathbb{R} \times \mathrm{cl}(Z) \to \mathrm{cl}(Z)$ by

$$\psi(t,(x,y)) = \begin{cases} g^{-1}\left(\phi^t(g(x,y))\right) & \text{if } x \in I, \\ (-1, y-t) & \text{if } x = -1, \\ (1, y+t) & \text{if } x = 1. \end{cases}$$

Prove that ψ is the flow corresponding to the system of differential equations

$$\dot{x} = \begin{cases} \dfrac{(1-x^2)^2[x - y(1-x^2)]}{(1+x^2)\left[(1-x^2) + \sqrt{x^2 + y^2(1-x^2)^2}\right]} & \text{if } x \in I, \\ 0 & \text{if } x = \pm 1, \end{cases}$$

$$\dot{y} = \begin{cases} \dfrac{x + y(1-x^2)}{1 - x^2 + \sqrt{x^2 + y^2(1-x^2)^2}} & \text{if } x \in I, \\ 1 & \text{if } x = 1, \\ -1 & \text{if } x = -1. \end{cases}$$

(d) Prove that if $p \in Z \setminus \{(0,0)\}$, then the ω-limit set of p with respect to ψ is neither compact nor connected. Compare with Proposition 2.3.6 (vi).

(11) Consider the flow ϕ^t corresponding to the differential equation

$$\dot{x} = \begin{pmatrix} 0 & -1 & 0 & 0 \\ 1 & 0 & 0 & 0 \\ 0 & 0 & 0 & -\frac{\sqrt{2}}{2} \\ 0 & 0 & \frac{\sqrt{2}}{2} & 0 \end{pmatrix} x$$

on $\mathbb{R}^4$. Show that there is a point $x \in \mathbb{R}^4$ such that $\mathcal{O}(x) \subseteq \omega(x)$ and x is not a periodic point of ϕ^t.

(12) Provide an example of a flow ϕ^t on the 3-sphere $S^3 = \{(x,y,z,w) \in \mathbb{R}^4 \mid x^2 + y^2 + z^2 + w^2 = 1\}$ such that the set of Poincaré recurrent points of ϕ^t is not closed. *Hint:* Reparametrize the flow in Example 2.4.6 so that it extends to the one-point compactification of $\mathbb{R}^3$. Map the extended flow to S^3 via the inverse of stereographic projection.

(13) Let ϕ^t be a flow on a topological space. Let T_0 be a positive real number. Prove that $x \in \Omega(\phi^t)$ if and only if for each open

neighborhood U of x there exists a real number $T \geq T_0$ such that $U \cap \phi^T(U) \neq \emptyset$.

(14) Prove or disprove: If ϕ^t is a flow on a second-countable Hausdorff topological space, then $C(\phi^t) = \mathrm{cl}(\mathrm{Rec}(\phi^t))$.

(15) Prove that if ϕ^t is a flow on a compact metric space X and $x \in X$, then $x \in \Omega^+(x)$ if and only if $x \in \Omega^-(x)$.

(16) Let $H = \{(x, y) \in \mathbb{R}^2 \mid y > 0\}$. Define $\phi : \mathbb{R} \times H \to H$ by

$$\phi(t, (x, y)) = (x + t, y).$$

Let d_E be the metric on H induced by the standard metric on $\mathbb{R}^2$. Let d_H be the metric on H induced by the hyperbolic metric on $\mathbb{H}^2$. See Example 2.7.13.

(a) Prove that ϕ is a flow.

(b) Prove that the identity map $\iota : (H, d_E) \to (H, d_H)$ is a topological conjugacy from ϕ^t to itself. *Hint:* The metrics d_E and d_H induce the same topology on H.

(c) Prove that the chain recurrent set of ϕ^t with respect to d_E is the empty set. *Hint:* See Example 2.7.7.

(17) Let ϕ^t be the flow on $\mathbb{R}$ defined by

$$\phi^t(x) = x + t.$$

Let ψ^t be the flow on $S^1 \setminus \{0\}$ defined by

$$\psi^t(\theta) = \pi - 2\arctan\left[\tan\left(\frac{\pi - \theta}{2}\right) + t\right].$$

(a) Prove that ϕ^t is topologically conjugate to ψ^t via a homeomorphism $h : \mathbb{R} \to S^1 \setminus \{0\}$.

(b) Prove that $\mathcal{R}(\psi^t) \neq h(\mathcal{R}(\phi^t))$.

(18) Let ϕ^t be a flow on a Hausdorff topological space X. Define the **Newhouse limit set** of ϕ^t by

$$L(\phi^t) = \bigcup_{x \in X} (\alpha(x) \cup \omega(x)).$$

(a) Prove that if X is first-countable, then $L(\phi^t)$ is an invariant set.

(b) Provide an example of a flow ϕ^t such that $L(\phi^t)$ is not closed.

(c) Let Y be a Hausdorff topological space. Prove that if (h, ρ) is a topological equivalence from ϕ^t to a flow $\psi^t : Y \to Y$, then

$$L(\psi^t) = h(L(\phi^t)).$$

(d) Prove that $\alpha(x) \cup \omega(x) \subseteq L(\phi^t)$ for all $x \in X$.

(e) Provide an example of a flow ϕ^t such that $L(\phi^t|_{L(\phi^t)}) \neq L(\phi^t)$.

(f) Prove that if X is second-countable, then $C(\phi^t) \subseteq L(\phi^t) \subseteq \Omega(\phi^t)$. Provide examples which show that each containment may be proper.

Chapter 3

Irreducible Sets

One of our objectives is to separate the phase space of a flow into two invariant subsets, one in which every point is recurrent and another in which no point is recurrent. Furthermore, we seek to partition the collection of recurrent points into closed invariant subsets which are dynamically irreducible in some sense. The goal of this chapter is to develop types of irreducibility.

3.1. Minimal Sets

Our first notion of irreducibility is minimality. Birkhoff [**6**] introduced minimality.

Definition 3.1.1. Let ϕ^t be a flow on a topological space X. A subset A of X is **minimal** with respect to ϕ^t if A is nonempty, A is closed in X, and $\mathcal{O}(x)$ is dense in A with respect to the topology on X for all $x \in A$. Denote the union of all minimal sets with respect to ϕ^t by $M(\phi^t)$.

Equivalently, a nonempty set A is minimal if $A = \mathrm{cl}(\mathcal{O}(x))$ for all $x \in A$.

Example 3.1.2. A periodic orbit is a minimal set. In particular, a set consisting of a single fixed point is a minimal set. Therefore,

$$\mathrm{Fix}(\phi^t) \subseteq \mathrm{Per}(\phi^t) \subseteq M(\phi^t)$$

for any flow ϕ^t.

A subset of a Hausdorff space which contains a fixed point and has more than one element is not minimal.

Example 3.1.3. Let ϕ^t be the flow corresponding to the constant differential equation

$$\dot{x} = 1$$

on $\mathbb{R}$. The flow ϕ^t has one orbit, the entire real line. Thus, $\mathbb{R}$ is a minimal set with respect to ϕ^t.

Example 3.1.4. Consider an irrational flow on the torus $\mathbb{T}^2$. Since $\mathbb{T}^2$ is nonempty and closed, the entire torus $\mathbb{T}^2$ is a minimal set with respect to an irrational flow by Example 2.3.10.

With respect to an irrational flow the torus has no proper nonempty closed invariant subsets, illustrating an important general fact: a nonempty closed invariant set A is minimal if and only if A has no proper nonempty closed invariant subsets.

Proposition 3.1.5. *Let ϕ^t be a flow on a topological space X. A subset A of X is minimal with respect to ϕ^t if and only if A is nonempty, A is closed, A is invariant with respect to ϕ^t, and A has no proper nonempty closed invariant subsets with respect to ϕ^t.*

Proof. If A is a minimal subset of X with respect to ϕ^t, then A is nonempty and closed in X by definition. Let S be a nonempty closed invariant subset of A. Since S is nonempty, there exists $x \in S$. Because the orbit of each point in A is dense in A with respect to the topology on X, and S is a closed invariant subset of A,

$$A = \mathrm{cl}(\mathcal{O}(x)) \subseteq S \subseteq A.$$

Thus, $S = A$. Therefore, A has no proper nonempty closed invariant subsets. Furthermore, A is invariant since S is invariant.

Conversely, let A be a nonempty closed invariant subset of X such that A has no proper nonempty closed invariant subsets. If $x \in A$, then the invariance of A guarantees that $\mathcal{O}(x) \subseteq A$. Proposition 1.4.5 implies that $\mathrm{cl}(\mathcal{O}(x))$ is invariant. Thus, $\mathrm{cl}(\mathcal{O}(x))$ is a nonempty closed invariant subset of A. Since A has no proper nonempty closed invariant subsets, $A = \mathrm{cl}(\mathcal{O}(x))$. Hence, $\mathcal{O}(x)$ is dense in A with respect to the topology on X for all $x \in A$. Therefore, A is a minimal set with respect to ϕ^t. $\square$

Proposition 3.1.5 shows that a minimal set is irreducible in the sense that it does not contain any proper nonempty closed invariant subsets. Some texts define a minimal set to be a nonempty closed invariant set which does not have a proper nonempty closed invariant subset. We define minimality in terms of the density of orbits so that its definition more closely parallels the definition of topological transitivity in Section 3.2 and the definition of chain transitivity in Section 3.3.

The minimal sets of a flow are connected.

Proposition 3.1.6. *A minimal set of a flow is connected.*

Proof. Let ϕ^t be a flow on a topological space X. Let A be a minimal set with respect to ϕ^t, and let $x \in A$. By Proposition 1.1.11 the set $\mathcal{O}(x)$ is connected. Because the closure of a connected set is connected, $\mathrm{cl}(\mathcal{O}(x))$ is a connected set. Since A is minimal, the orbit of x is dense in A with respect to the topology on X. So, $A = \mathrm{cl}(\mathcal{O}(x))$. Therefore, A is connected. □

Compact minimal sets have a characterization via α- and ω-limits sets.

Proposition 3.1.7. *Let ϕ^t be a flow on a first-countable Hausdorff topological space X.*

(i) *A nonempty compact subset A of X is minimal with respect to ϕ^t if and only if $\omega(x) = A$ for all $x \in A$*

(ii) *A nonempty compact subset A of X is minimal with respect to ϕ^t if and only if $\alpha(x) = A$ for all $x \in A$.*

Proof. (i) Let A be a minimal set with respect to ϕ^t, and let $x \in A$. Because A is compact and minimal, $\omega(x)$ is a nonempty closed invariant subset of A by Proposition 2.3.6 (v) and (vi). The minimality of A and Proposition 3.1.5 imply that $\omega(x) = A$.

Conversely, suppose that $\omega(x) = A$ for all $x \in A$. If B is a nonempty closed invariant subset of A and $x \in B$, then $A = \omega(x) \subseteq B \subseteq A$; so $A = B$. Therefore, A is a minimal set with respect to ϕ^t by Proposition 3.1.5.

(ii) The proof is similar to the proof of part (i). □

Every element of a minimal set is Poincaré recurrent.

Proposition 3.1.8. *The Poincaré recurrent set of a flow on a compact first-countable Hausdorff topological space contains every minimal set of the flow.*

Proof. Let ϕ^t be a flow on a compact first-countable Hausdorff topological space. Let A be a minimal set with respect to ϕ^t with $x \in A$. By Proposition 3.1.7, $\alpha(x) = \omega(x) = A$. So, $x \in \omega(x) \cap \alpha(x)$. Therefore, x is Poincaré recurrent. Consequently, $M(\phi^t) \subseteq \mathrm{Rec}(\phi^t)$. □

While a flow (such as an irrational flow on the torus) may have an empty set of periodic points, a flow on a nonempty compact space always has a minimal set.

Proposition 3.1.9. *Every flow on a nonempty compact topological space has a minimal set.*

Proof. Let ϕ^t be a flow on a nonempty compact topological space X. Let $\mathcal{C}$ be the collection of nonempty closed invariant subsets of X partially ordered by inclusion. The collection $\mathcal{C}$ is nonempty because $X \in \mathcal{C}$. Since X is compact, intersections of nested nonempty closed invariant sets are nonempty, closed and invariant. Thus, every totally ordered subset of $\mathcal{C}$ has a lower bound. By Zorn's Lemma, $\mathcal{C}$ has a minimal element A. That is, A is a nonempty closed invariant subset of X which has no proper nonempty closed invariant subsets. Therefore, A is a minimal set with respect to ϕ^t by Proposition 3.1.5. □

To summarize, we have the following inclusions for a flow ϕ^t on a second-countable metric space X:

$$\mathrm{Fix}(\phi^t) \subseteq \mathrm{Per}(\phi^t) \subseteq M(\phi^t) \subseteq \mathrm{Rec}(\phi^t) \subseteq C(\phi^t) \subseteq \Omega(\phi^t) \subseteq \mathcal{R}(\phi^t).$$

Proposition 3.1.9 guarantees that if X is nonempty and compact, then $M(\phi^t)$ is nonempty, so that $\mathrm{Rec}(\phi^t)$, $C(\phi^t)$, $\Omega(\phi^t)$ and $\mathcal{R}(\phi^t)$ are nonempty as well.

We conclude our discussion of minimality by showing that topological equivalences carry minimal sets to minimal sets.

Proposition 3.1.10. *Let X and Y be topological spaces. If (h, ρ) is a topological equivalence from the flow $\phi^t : X \to X$ to the flow $\psi^t : Y \to Y$, and $A \subseteq X$ is minimal with respect to ϕ^t, then $h(A)$ is minimal with respect to ψ^t.*

Proof. The set $h(A)$ is nonempty and closed because h is a homeomorphism. By Proposition 1.4.6, the set $h(A)$ is invariant with respect to ψ^t. We shall prove that $h(A)$ has no nonempty proper closed invariant subsets. Let B be a nonempty closed invariant subset of $h(A)$. Since h is a homeomorphism, $h^{-1}(B)$ is a nonempty closed subset of A. Furthermore, $h^{-1}(B)$ is invariant by Proposition 1.4.6. Thus, $h^{-1}(B)$ is a nonempty closed invariant subset of A. Since A is minimal with respect to ϕ^t, Proposition 3.1.5 implies that $h^{-1}(B) = A$. Consequently, $B = h(h^{-1}(B)) = h(A)$. So, $h(A)$ has no nonempty proper closed invariant subsets. Therefore, $h(A)$ is minimal with respect to ψ^t by Proposition 3.1.5 . □

Denjoy [**12**] constructs a flow on the torus with a complicated minimal set. Schwartz [**39**] classifies minimal sets for C^2 flows on compact surfaces. Ellis [**13**] explores the classification of minimal sets for abstract dynamical systems.

3.2. Topologically Transitive Sets

Section 3.1 examined minimality as one type of dynamical irreducibility for an invariant set. A nonempty set A is minimal if the orbit of every point in A is dense in A with respect to the topology of the phase space. In this section we require the existence of only a single point whose forward and backward orbits are each dense in A. Corollary 3.2.15 more fully relates this condition to the irreducibility of an invariant set.

Definition 3.2.1. Let ϕ^t be a flow on a topological space X. A set $A \subseteq X$ is **topologically transitive** with respect to ϕ^t if there exists $x \in A$ such that $\mathcal{O}^+(x)$ and $\mathcal{O}^-(x)$ are dense in A with respect to the topology on X.

It is common to apply the term topologically transitive to a flow $\phi^t : X \to X$ rather than to a set. In our terminology, this means that X is topologically transitive with respect to ϕ^t. We prefer to place the emphasis on the set rather than the flow.

To provide intuition for topological transitivity, let A be a topologically transitive set with respect to a flow $\phi^t : X \to X$. By definition there exists $x \in A$ such that $\mathcal{O}^+(x)$ and $\mathcal{O}^-(x)$ are dense in A. If $y, z \in A$, then for each open neighborhood U of y in X and each open neighborhood V of z in X there exist nonnegative real numbers r and s such that $\phi^{-r}(x) \in U$ and $\phi^s(x) \in V$. Thus, we can hop within U from y to $\phi^{-r}(x)$, follow the orbit of x for time $r+s$ and hop within V from $\phi^s(x)$ to z. That is, we can "transit" from any point $y \in A$ to any other point $z \in A$ by catching the orbit of x in any neighborhood of y and riding the orbit of x in forward time to any neighborhood of z.

Example 3.2.2. If p is a periodic point of a flow ϕ^t, then the orbit of p is topologically transitive with respect to ϕ^t. In particular, a singleton consisting of a fixed point is topologically transitive.

Example 3.2.3. Consider the decreasing segment flow of Example 2.7.7 on an interval $[a, b]$. The points a and b are fixed points. Thus, their forward orbits are not dense in $[a, b]$. If $x \in (a, b)$, then the forward orbit of x is $[x, b)$ and the backward orbit of x is $(a, x]$ neither of which is dense in $[a, b]$. Therefore, $[a, b]$ is not topologically transitive with respect to ϕ^t. Notice that the orbit of x is (a, b) and is dense in $[a, b]$.

It is common to apply the term topologically transitive to a flow $\phi^t : X \to X$ for which there exists $x \in X$ such that the orbit of x is dense in X. This is Birkhoff's original notion of topological transitivity. See Birkhoff [**7**]. The decreasing segment flow in Example 3.2.3 illustrates how such a definition may not capture the dynamical irreducibility of an invariant set. In that example, the invariant set $[a, b]$ decomposes into the two singleton fixed points $\{a\}$ and $\{b\}$ connected by the orbit (a, b). Yet, the orbit of every point in (a, b) is dense in $[a, b]$.

The following example shows that topological transitivity in the sense of Definition 3.2.1 does not generalize minimality. However, Corollary 3.2.13 will show that every compact minimal set is topologically transitive.

Example 3.2.4. Consider the flow ϕ^t associated with the constant differential equation

$$\dot{x} = 1$$

on $\mathbb{R}$. By Example 3.1.3, the set $\mathbb{R}$ is minimal with respect to ϕ^t. However, if $x \in \mathbb{R}$, then $\mathcal{O}^+(x) = [x, \infty)$ is not dense in $\mathbb{R}$. Therefore, $\mathbb{R}$ is not topologically transitive with respect to ϕ^t in the sense of Definition 3.2.1.

To provide an example of a topologically transitive set which is not minimal we modify an irrational flow on the torus.

Example 3.2.5. Let c_1 and c_2 be real numbers such that c_1/c_2 is irrational. Consider the irrational flow ϕ^t corresponding to the system of differential equations

$$\begin{aligned} \dot{x}_1 &= c_1, \\ \dot{x}_2 &= c_2 \end{aligned}$$

on $\mathbb{T}^2$. Example 2.3.10 demonstrated that the forward and backward orbits of a point under an irrational flow are dense in the torus. Therefore, the torus is topologically transitive with respect to an irrational flow.

Let $p \in \mathbb{T}^2$. Let $\beta : \mathbb{T}^2 \to \mathbb{R}$ be a smooth nonnegative function such that $\beta^{-1}(\{0\}) = \{p\}$, and

$$\int_0^t \beta(\phi^\tau(x))\, d\tau \to \pm\infty$$

as $t \to \pm\infty$, respectively, for all $x \in \mathbb{T}^2 \setminus \mathcal{O}(p, \phi^t)$.

Consider the flow $\psi^t : \mathbb{T}^2 \to \mathbb{T}^2$ corresponding to the system of differential equations

$$\begin{aligned} \dot{x}_1 &= c_1\beta(x), \\ \dot{x}_2 &= c_2\beta(x) \end{aligned}$$

on the torus $\mathbb{T}^2$. Since $\beta(p) = 0$, the point p is a fixed point of ψ^t. So, $\{p\}$ is a proper, nonempty, closed invariant subset of the torus with respect to ψ^t. Thus, $\mathbb{T}^2$ is not minimal with respect to ψ^t.

Let $A = \mathbb{T}^2 \setminus \mathcal{O}(p, \phi^t)$. Since A is the complement of the invariant set $\mathcal{O}(p, \phi^t)$ with respect to ϕ^t, the set A is invariant with respect to ϕ^t by Proposition 1.4.4 (i). Define $\rho : \mathbb{R} \times A \to A$ by

$$\rho(t, x) = \int_0^t \beta(\phi^\tau(x))\, d\tau.$$

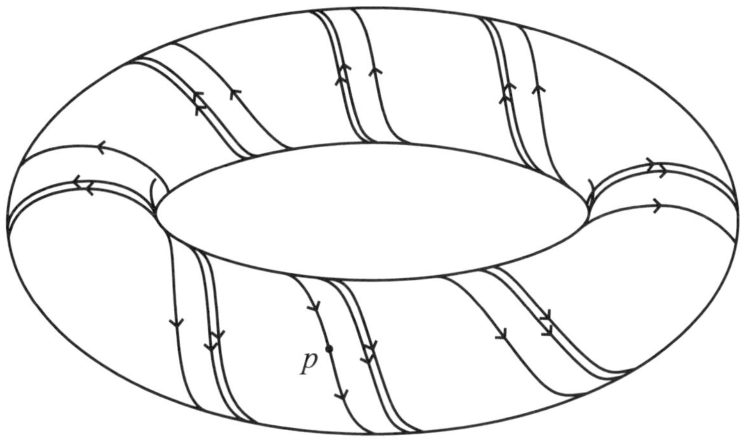

Figure 3.1. A phase portrait of the flow of Example 3.2.5

If $\iota : A \to A$ is the restriction to A of the identity map on $\mathbb{T}^2$, then (ι, ρ) is a topological equivalence from $\phi^t|_A$ to $\psi^t|_A$ by Example 1.3.10. Since $\rho(0, x) = 0$ for all $x \in A$, we have $\mathcal{O}^{\pm}(x, \psi^t) = \mathcal{O}^{\pm}(x, \phi^t)$ for all $x \in A$. Consequently, if $x \in A$, then $\mathrm{cl}(\mathcal{O}^+(x, \psi^t)) = \mathrm{cl}(\mathcal{O}^+(x, \phi^t)) = \mathbb{T}^2$. Hence, $\mathcal{O}^+(x, \psi^t)$ is dense in $\mathbb{T}^2$. Similarly, $\mathcal{O}^-(x, \psi^t)$ is dense in $\mathbb{T}^2$. Therefore, the torus $\mathbb{T}^2$ is topologically transitive, but not minimal, with respect to ψ^t. See Figure 3.1.

The fact that a topologically transitive set is closed and invariant follows directly from Definition 3.2.1.

Proposition 3.2.6. *Every topologically transitive set is nonempty, closed and invariant.*

Proof. If A is a topologically transitive set, then there exists $x \in A$ such that $\mathcal{O}^+(x)$ and $\mathcal{O}^-(x)$ are dense in A. In particular, A is nonempty, and $A = \mathrm{cl}(\mathcal{O}^+(x)) = \mathrm{cl}(\mathcal{O}^-(x))$. Thus, A is closed. Furthermore,

$$A = A \cup A = \mathrm{cl}(\mathcal{O}^+(x)) \cup \mathrm{cl}(\mathcal{O}^-(x)) = \mathrm{cl}(\mathcal{O}^+(x) \cup \mathcal{O}^-(x)) = \mathrm{cl}(\mathcal{O}(x)).$$

Since the orbit of x is invariant, A is invariant by Proposition 1.4.5. □

Next, we characterize topological transitivity in terms of open sets.

Proposition 3.2.7. *Let ϕ^t be flow on a topological space X. If $A \subseteq X$ is topologically transitive with respect to ϕ^t, then for all open sets U and V in X such that $U \cap A$ and $V \cap A$ are nonempty, there exists a nonnegative real number T such that $\phi^T(U) \cap V$ is nonempty.*

Proof. Let U and V be open sets in X such that $U \cap A$ and $V \cap A$ are nonempty. Since A is topologically transitive, there exists $x \in A$ such that $\mathcal{O}^+(x)$ and $\mathcal{O}^-(x)$ are dense in A with respect to the topology on X. So, there exist nonnegative real numbers r and s such that $\phi^{-r}(x) \in U$ and $\phi^s(x) \in V$. Let $T = r + s$. The number T is nonnegative, and $\phi^s(x) \in \phi^s(\phi^r(U)) \cap V = \phi^T(U) \cap V$. Therefore, $\phi^T(U) \cap V$ is nonempty. □

By restricting the spaces under consideration, we can obtain a converse to Proposition 3.2.7. The Birkhoff Transitivity Theorem is the key result in obtaining such a converse. Recall two definitions.

Definition 3.2.8. A subset Y of a topological space X is called **residual** in X if Y is an intersection of a countable collection of open subsets in X each of which is dense in X.

Definition 3.2.9. A topological space X is called a **Baire space** if every subset of X which is residual in X is dense in X.

A topological space X is a Baire space if the intersection of every countable collection of open dense subsets of X is dense in X. Notice that a residual subset of a nonempty Baire space is nonempty.

Theorem 3.2.10. The Birkhoff Transitivity Theorem. *Let ϕ^t be a flow on a nonempty second-countable Baire space X.*

(i) *If $\bigcup_{t\geq 0} \phi^t(U)$ is dense in X for every open subset U of X, then there exists $\mathcal{D}^- \subseteq X$ such that $\mathcal{D}^-$ is residual in X, and $\mathcal{O}^-(x)$ is dense in X for all $x \in \mathcal{D}^-$.*

(ii) *If $\bigcup_{t\leq 0} \phi^t(U)$ is dense in X for every open subset U of X, then there exists $\mathcal{D}^+ \subseteq X$ such that $\mathcal{D}^+$ is residual in X, and $\mathcal{O}^+(x)$ is dense in X for all $x \in \mathcal{D}^+$.*

(iii) *If $\bigcup_{t\geq 0} \phi^t(U)$ and $\bigcup_{t\leq 0} \phi^t(U)$ are dense in X for every open subset U of X, then there exists $\mathcal{D} \subseteq X$ such that $\mathcal{D}$ is residual in X, and $\mathcal{O}^+(x)$ and $\mathcal{O}^-(x)$ are dense in X for all $x \in \mathcal{D}$.*

Proof. (i) Let $\{V_1, V_2, V_3, \ldots\}$ be a countable base for the topology of X. The sets $V_1, V_2, V_3, \ldots$ are open in X. By hypothesis, $\bigcup_{t\geq 0} \phi^t(V_i)$ is dense in X for each $i = 1, 2, 3, \ldots$. Furthermore, for each real number t the map ϕ^t is a homeomorphism. So, $\phi^t(V_i)$ is open in X for each $i = 1, 2, 3, \ldots$. Thus, $\bigcup_{t\geq 0} \phi^t(V_i)$ is open in X for each $i = 1, 2, 3, \ldots$. The set

$$\mathcal{D}^- = \bigcap_{i=1}^{\infty} \bigcup_{t\geq 0} \phi^t(V_i)$$

is an intersection of countably many dense open subsets of X. Therefore, $\mathcal{D}^-$ is residual in X.

Since X is a nonempty Baire space, $\mathcal{D}^-$ is nonempty. If $x \in \mathcal{D}^-$, then $x \in \bigcup_{t\geq 0} \phi^t(V_i)$ for all $i = 1, 2, 3, \ldots$. For each $i = 1, 2, 3, \ldots$ there exists $t \geq 0$ such that $x \in \phi^t(V_i)$. That is, $\phi^{-t}(x) \in V_i$. Thus, $\mathcal{O}^-(x) \cap V_i$ is nonempty for all $i = 1, 2, 3, \ldots$. Since $\{V_1, V_2, V_3, \ldots\}$ is a base for the topology of X, the backward orbit of x is dense in X.

(ii) The proof is similar to the proof of part (i).

(iii) The set $\mathcal{D} = \mathcal{D}^+ \cap \mathcal{D}^-$ is residual in X, and $\mathcal{O}^+(x)$ and $\mathcal{O}^-(x)$ are dense in X for all $x \in \mathcal{D}$ by parts (i) and (ii). □

From the Birkhoff Transitivity Theorem we obtain a converse to Proposition 3.2.7 on second-countable Baire spaces.

Proposition 3.2.11. *Let ϕ^t be flow on a second-countable Baire space X. If for each pair of open sets U and V in X such that $U \cap A$ and $V \cap A$ are nonempty there exists a nonnegative real number T such that $\phi^T(U) \cap V$ is nonempty, then A is topologically transitive with respect to ϕ^t.*

Proof. Let U be an open subset in X. For every open set V in X, there exists a nonnegative real number T such that $\phi^T(U) \cap V$ is nonempty. Thus, $\bigcup_{t\geq 0} \phi^t(U)$ is dense in A. Similarly, there exists a nonnegative real number S such that $\phi^S(V) \cap U$ is nonempty. So, $\phi^{-S}(U) \cap V$ is nonempty. Thus, $\bigcup_{t\leq 0} \phi^t(U)$ is dense in A. By the Birkhoff Transitivity Theorem, there exists $\mathcal{D} \subseteq A$ such that $\mathcal{D}$ is residual, and $\mathcal{O}^+(x)$ and $\mathcal{O}^-(x)$ are dense in A for all $x \in \mathcal{D}$. Since X is a nonempty Baire space, $\mathcal{D}$ is nonempty. Therefore, A is topologically transitive with respect to ϕ^t. □

Next, we characterize topological transitivity in terms of limit sets and prove that every compact minimal set is topologically transitive.

Proposition 3.2.12. *Let ϕ^t be a flow on a first-countable Hausdorff topological space X. A set $A \subseteq X$ is topologically transitive with respect to ϕ^t if and only if $A = \alpha(x) = \omega(x)$ for some $x \in A$.*

Proof. If $A = \alpha(x) = \omega(x)$ for some $x \in A$, then both the forward and backward orbits of x are dense in A. Thus, A is topologically transitive with respect to ϕ^t.

Conversely, let A be topologically transitive with respect to ϕ^t. Then $A = \mathrm{cl}\,(\mathcal{O}^+(x))$ for some $x \in A$. Let $y \in A$. We will show that $y \in \omega(x)$.

Since X is first-countable, there is a countable base $\{U_k\}_{k=1}^{\infty}$ for the topology of X at y. For each positive integer n, let

$$B_n = \bigcap_{k=1}^{n} U_k.$$

For each positive integer n, the set $B_n \setminus \{\phi^t(x) \,|\, 0 \le t \le n\}$ is open in X.

If $B_n \setminus \{\phi^t(x) \,|\, 0 \le t \le n\}$ is nonempty for each positive integer n, then since the forward orbit of x is dense in A, there exists a sequence of real numbers $\{t_n\}$ so that $\phi^{t_n}(x) \in B_n \setminus \{\phi^t(x) \,|\, 0 \le t \le n\}$ for each positive integer n. In particular, $t_n \ge n$, so $\lim_{n\to\infty} t_n = +\infty$. We claim that $\lim_{n\to\infty} \phi^{t_n}(x) = y$. If V is an open set in X containing y, then there is a positive integer N such that $U_N \subseteq V$. Then $B_n \subseteq V$ for each $n \ge N$. Hence, $\phi^{t_n}(x) \in V$. So, $\lim_{n\to\infty} \phi^{t_n}(x) = y$. Thus, $y \in \omega(x)$.

If there exists a positive integer n such that $B_n \setminus \{\phi^t(x) \,|\, 0 \le t \le n\}$ is empty, then $B_n \subseteq \{\phi^t(x) \,|\, 0 \le t \le n\}$. Since $\mathcal{O}^-(x)$ is dense in A, there exists a positive real number r such that $\phi^{-r}(x) \in B_n \setminus \{x\}$. Because $B_n \subseteq \{\phi^t(x) \,|\, 0 \le t \le n\}$, there exists $s \in [0, n]$ such that $\phi^{-r}(x) = \phi^s(x)$. Thus, $r+s > 0$ and $\phi^{r+s}(x) = x$. Consequently, x is periodic, and $A = \mathrm{cl}(\mathcal{O}^+(x)) = \mathcal{O}(x) = \omega(x)$. The argument that $y \in \alpha(x)$ is similar. Therefore, $A = \alpha(x) = \omega(x)$. □

Corollary 3.2.13. *Every compact minimal set with respect to a flow on a first-countable Hausdorff topological space is topologically transitive with respect to the flow.*

Proof. Let A be a compact minimal set with respect to a flow on a first-countable Hausdorff topological space. By Proposition 3.1.7, $A = \omega(x) = \alpha(x)$ for all $x \in A$. Therefore, by Proposition 3.2.12, A is topologically transitive. □

Like minimal sets, topologically transitive sets for flows are connected.

Proposition 3.2.14. *A topologically transitive set with respect to a flow is connected.*

Proof. If A is a topologically transitive set with respect to a flow, then $A = \mathrm{cl}\,(\mathcal{O}^+(x))$ for some $x \in A$. By Proposition 1.1.11, $\mathcal{O}^+(x)$ is connected. Therefore, $A = \mathrm{cl}\,(\mathcal{O}^+(x))$ is connected. □

A topologically transitive set does not partition into nonempty closed subsets. In particular, it does not partition into nonempty closed invariant subsets.

Corollary 3.2.15. *If A is a topologically transitive set, then there do not exist nonempty disjoint closed subsets B and C of A such that $A = B \cup C$.*

Proof. If A is a topologically transitive set, then A is connected by Proposition 3.2.14. Therefore, there do not exist nonempty disjoint closed subsets B and C of A such that $A = B \cup C$. $\square$

Finally, topological equivalences carry topologically transitive sets to topologically transitive sets.

Proposition 3.2.16. *Let X and Y be topological spaces. If (h, ρ) is a topological equivalence from the flow $\phi^t : X \to X$ to the flow $\psi^t : Y \to Y$, and $A \subseteq X$ is topologically transitive with respect to ϕ^t, then $h(A)$ is topologically transitive with respect to ψ^t.*

Proof. Since A is topologically transitive there exists $x \in A$ such that $\mathcal{O}^+(x, \phi^t)$ is dense in A. Reparametrizations are surjective, so there exists a real number τ such that $\rho(\tau, x) = 0$. Let $y = \psi^\tau(h(x))$. Because (h, ρ) is a topological equivalence, $\mathcal{O}^+(y, \psi^t) = h(\mathcal{O}^+(x, \phi^t))$ is dense in $h(A)$. Similarly, $\mathcal{O}^-(y)$ is dense in $h(A)$. Therefore, $h(A)$ is topologically transitive with respect to ψ^t. $\square$

Smale's Spectral Decomposition Theorem says that for certain flows (called Axiom A) the nonwandering set partitions into nonempty closed invariant sets each of which is topologically transitive. See Smale [**41**]. This establishes the Decomposition Property of the nonwandering set for Axiom A flows.

3.3. Chain Transitive Sets

In this section we use (ϵ, T)-chains to extend topological transitivity to a more general type of dynamical irreducibility due to Conley [**11**] called chain transitivity.

By Definition 2.7.1 two points x and y are chain equivalent if for each $\epsilon > 0$ and $T > 0$ there exists an (ϵ, T)-chain from x to y and there exists an (ϵ, T)-chain from y to x. We examine how the chain equivalence relation partitions the chain recurrent set into closed invariant maximally chain transitive subsets called chain components. We prove that the chain components of a flow on a compact metric space are exactly the connected components of the chain recurrent set and that the chain recurrent set enjoys the Restriction Property.

Definition 3.3.1. Let ϕ^t be a flow on a metric space X. A set $A \subseteq X$ is **chain transitive** with respect to ϕ^t if A is a nonempty closed invariant set with respect to ϕ^t such that for each $x, y \in A$, $\epsilon > 0$ and $T > 0$ there exists an (ϵ, T)-chain from x to y.

Intuitively, within a chain transitive set it is possible to "transit" from any point x to any point y by riding along an (ϵ, T)-chain for any $\epsilon > 0$ and $T > 0$. By choosing $y = x$, we see that every element of a chain transitive set is chain recurrent. Therefore, the chain recurrent set of a flow contains every set which is chain transitive with respect to the flow.

The following proposition shows that every topologically transitive set is chain transitive.

Proposition 3.3.2. *Every topologically transitive set with respect to a flow on a metric space is chain transitive with respect to the flow.*

Proof. Let A be a topologically transitive set with respect to a flow ϕ^t on a metric space (X, d). The set A is nonempty, closed and invariant with respect to ϕ^t by Proposition 3.2.6.

By Proposition 3.2.12 there exists $z \in A$ such that $A = \omega(z)$. Let $x, y \in A$, and let $\epsilon > 0$ and $T > 0$. By the continuity of ϕ^T there exists $\delta > 0$ such that if $w \in X$ and $d(w, x) < \delta$, then $d(\phi^T(w), \phi^T(x)) < \epsilon$. Since $x, y \in \omega(z)$, there exists $w \in \mathcal{O}^+(z)$ such that $d(w, x) < \delta$, and there exists $S > 2T$ such that $d(\phi^S(w), y) < \epsilon$. Thus,

$$(x, \phi^T(w), y; T, S - T)$$

is an (ϵ, T)-chain from x to y. Therefore, A is chain transitive with respect to ϕ^t. □

However, as the following example demonstrates, a set may be chain transitive but not topologically transitive.

Example 3.3.3. The arrested rotation ϕ^t corresponds to the differential equation

$$\dot{\theta} = \sin^2 \theta$$

on S^1. By Example 2.3.5 the ω-limit set of each element of S^1 consists of a single fixed point. Therefore, S^1 is not topologically transitive with respect to ϕ^t by Proposition 3.2.12.

We claim that S^1 is chain transitive with respect to ϕ^t. Clearly, S^1 is nonempty, closed and invariant with respect to ϕ^t. Let $x, y \in S^1$, and let $\epsilon > 0$ and $T > 0$. Denote the usual metric on S^1 by d.

Case I. Assume that $0 < x \bmod 2\pi \leq \pi$ and $\pi < y \bmod 2\pi \leq 2\pi$. By Example 2.3.5, $\omega(x) = \{\pi\}$. So, there exists $t_0 \in [T, \infty)$ such that $d(\phi^{t_0}(x), \pi) < \epsilon/2$. Again by Example 2.3.5, $\alpha(y) = \{\pi\}$. So, there exist $x_1 \in S^1$ and $t_1 \in [T, \infty)$ such that $\pi < x_1 \bmod 2\pi < \pi + \epsilon/2$ and $d(\phi^{t_1}(x_1), y) = 0 < \epsilon$. By the triangle inequality,

$$d(\phi^{t_0}(x), x_1) < \epsilon.$$

Therefore, $(x, x_1, y; t_0, t_1)$ is an (ϵ, T)-chain from x to y.

Case II. Assume that $0 < y \bmod 2\pi \leq \pi$ and $\pi < x \bmod 2\pi \leq 2\pi$. The construction is similar to that of Case I.

Case III. Assume that $0 < x \bmod 2\pi \leq \pi$ and $0 < y \bmod 2\pi \leq \pi$. By Cases I and II there exist (ϵ, T)-chains from x to $\frac{3\pi}{2}$ and from $\frac{3\pi}{2}$ to y. Concatenating these chains provides an (ϵ, T)-chain from x to y.

Case IV. Assume that $\pi < x \bmod 2\pi \leq 2\pi$ and $\pi < y \bmod 2\pi \leq 2\pi$. The construction is similar to that of Case III.

Therefore, S^1 is chain transitive, but not topologically transitive, with respect to the arrested rotation.

Topological equivalences carry chain transitive sets to chain transitive sets under further hypotheses on the topological equivalence.

Proposition 3.3.4. *Let (X, d_X) and (Y, d_Y) be metric spaces. Let (h, ρ) be a topological equivalence from the flow $\phi^t : X \to X$ to the flow $\psi^t : Y \to Y$ such that $h : X \to Y$ is uniformly continuous and $\sup\{\rho(T, w) \mid w \in X\}$ exists for each $T > 0$. If A is chain transitive with respect to ϕ^t, then $h(A)$ is chain transitive with respect to ψ^t.*

Proof. Let $x, z \in A$. Let $\epsilon > 0$ and $T > 0$. We shall prove that there is an (ϵ, T)-chain from $h(x)$ to $h(z)$.

Because h is uniformly continuous on X, there exists $\delta > 0$ such that if $z_1, z_2 \in X$ and $d_X(z_1, z_2) < \delta$, then $d_Y(h(z_1), h(z_2)) < \epsilon$. Let $S = \sup\{\rho(T, w) \mid w \in X\}$. There exists a (δ, S)-chain

$$(x = x_0, x_1, \ldots, x_n = z; s_0, \ldots, s_{n-1})$$

from x to z. Let $y_i = h(x_i)$ for each $i = 0, \ldots, n$. Since ρ is a reparametrization, for each $i = 0. \ldots, n-1$ there exists t_i so that $s_i = \rho(t_i, x_i)$. We claim that

$$(h(x) = y_0, y_1, \ldots, y_n = h(z); t_0, \ldots, t_{n-1})$$

is an (ϵ, T)-chain from $h(x)$ to $h(z)$.

Since $s_i \geq S$ for $i = 0, \ldots, n-1$,

$$\rho(t_i, x_i) = s_i \geq S \geq \rho(T, x_i)$$

for $i = 0, \ldots, n-1$. Because $\rho(\cdot, x_i)$ is an increasing function, $t_i \geq T$ for $i = 0, \ldots, n-1$. Since $d_X(\phi^{s_i}(x_i), x_{i+1}) < \delta$ for $i = 0, \ldots, n-1$, we have

$$\begin{aligned} d_Y(\psi^{t_i}(y_i), y_{i+1}) &= d_Y(\psi^{t_i}(h(x_i)), h(x_{i+1})) \\ &= d_Y(h(\phi^{\rho(t_i, x_i)}(x_i)), h(x_{i+1})) \\ &= d_Y(h(\phi^{s_i}(x_i)), h(x_{i+1})) \\ &< \epsilon \end{aligned}$$

for $i = 0, \ldots, n-1$. Thus,

$$(h(x) = y_0, y_1, \ldots, y_n = h(z); t_0, \ldots, t_{n-1})$$

is an (ϵ, T)-chain from $h(x)$ to $h(z)$. Therefore, $h(A)$ is chain transitive with respect to ϕ^t. □

Next we investigate maximal chain transitive sets.

Proposition 3.3.5. *The relation* $\sim$ *defined by*

$$x \sim y \text{ if and only if } x \text{ is chain equivalent to } y$$

is an equivalence relation on the chain recurrent set of a flow on a metric space.

Proof. Let ϕ^t be a flow on a metric space. If x is a chain recurrent point with respect to ϕ^t, then for each $\epsilon > 0$ and $T > 0$ there exists an (ϵ, T)-chain from x to itself. Thus, $x \sim x$. So, the relation $\sim$ is reflexive.

If $x \sim y$, then for each $\epsilon > 0$ and $T > 0$ there exists an (ϵ, T)-chain from x to y and there exists an (ϵ, T)-chain from y to x. Thus, $y \sim x$. So, the relation $\sim$ is symmetric.

If $x \sim y$ and $y \sim z$, then for each $\epsilon > 0$ and $T > 0$ there exist (ϵ, T)-chains from x to y, from y to x, from y to z and from z to y. By concatenating the (ϵ, T)-chains from x to y and from y to z we obtain an (ϵ, T)-chain from x to z. By concatenating the (ϵ, T)-chains from z to y and from y to x we obtain an (ϵ, T)-chain from z to x. Thus, $x \sim z$. So, the relation $\sim$ is transitive. Therefore, the relation $\sim$ is an equivalence relation. □

Definition 3.3.6. Let ϕ^t be a flow on a metric space. The equivalence relation $\sim$ on the chain recurrent set of ϕ^t defined by

$$x \sim y \text{ if and only if } x \text{ is chain equivalent to } y$$

is the **chain equivalence relation** for ϕ^t. An equivalence class of the chain equivalence relation for ϕ^t is a **chain component** of ϕ^t.

We shall demonstrate that a chain component is a maximal chain transitive set. To that end the following proposition shows that every chain component is invariant.

Proposition 3.3.7. *Every chain component of a flow on a metric space is invariant with respect to the flow.*

Proof. Let C be a chain component of a flow ϕ^t on a metric space. Let $x \in C$ and $\tau \in \mathbb{R}$. We must prove that $\phi^\tau(x) \in C$. Let $\epsilon > 0$ and $T > 0$. We shall produce (ϵ, T)-chains from $\phi^\tau(x)$ to x and from x to $\phi^\tau(x)$.

Since x is chain recurrent, there exists an $(\epsilon, T + |\tau|)$-chain

$$(x = x_0, x_1, \dots, x_n = x; t_0, \dots, t_{n-1})$$

from x to itself. Hence,

$$(\phi^\tau(x), x_1, \dots, x_n = x; t_0 - \tau, t_1, \dots, t_{n-1})$$

is an (ϵ, T)-chain from $\phi^\tau(x)$ to x.

To obtain an (ϵ, T)-chain from x to $\phi^\tau(x)$, the continuity of ϕ^τ guarantees that there exists $\delta > 0$ such that if $d(x, y) < \delta$, then

$$d(\phi^\tau(x), \phi^\tau(y)) < \epsilon.$$

Since x is chain recurrent, there exists a $(\min\{\delta, \epsilon\}, T + |\tau|)$-chain

$$(x = x_0, x_1, \dots, x_n = x; t_0, \dots, t_{n-1})$$

from x to itself. Since $d(\phi^{t_{n-1}}(x_{n-1}), x) < \delta$ we obtain

$$d(\phi^{t_{n-1}+\tau}(x_{n-1}), \phi^\tau(x)) < \epsilon.$$

So,

$$(x = x_0, \dots, x_{n-1}, \phi^\tau(x); t_0, \dots, t_{n-2}, t_{n-1} + \tau)$$

is an (ϵ, T)-chain from x to $\phi^\tau(x)$.

Concatenating the (ϵ, T)-chain from $\phi^\tau(x)$ to x with the (ϵ, T)-chain from x to ϕ^τ provides an (ϵ, T)-chain from $\phi^\tau(x)$ to itself. Thus, $\phi^\tau(x)$ is a chain recurrent point and is chain equivalent to $x \in C$. Hence, $\phi^\tau(x) \in C$. Therefore, each chain component of ϕ^t is invariant with respect to ϕ^t. □

Equipped with Proposition 3.3.7 we fulfill the promise made in Section 2.7 to establish the flow invariance of the chain recurrent set.

Corollary 3.3.8. *The chain recurrent set of a flow on a metric space is invariant with respect to the flow.*

Proof. The chain recurrent set of a flow ϕ^t on a metric space is the union of the chain components of ϕ^t. By Proposition 3.3.7 each chain component is invariant. Every union of invariant sets is invariant by Proposition 1.4.4 (iii). Therefore, the chain recurrent set of ϕ^t is invariant with respect to ϕ^t. □

Not only are chain components invariant, they are closed as well.

Proposition 3.3.9. *Every chain component of a flow on a metric space is closed.*

Proof. Let C be a chain component of a flow ϕ^t on a metric space (X,d), and let $z \in \mathrm{cl}(C)$. We shall prove that $z \in C$. Let $y \in C$, and let $\epsilon > 0$ and $T > 0$. We shall produce (ϵ, T)-chains from z to y and from y to z.

The continuity of ϕ^T implies that there exists $\delta > 0$ such that if $d(x,z) < \delta$, then

$$d(\phi^T(x), \phi^T(z)) < \epsilon.$$

Let $\eta = \min\{\delta, \epsilon/2\}$. Since z is a limit point of C, there exists $x \in C$ such that $d(x,z) < \eta$. Because $x, y \in C$, there exists an $(\eta, 2T)$-chain

$$(x = x_0, x_1, \ldots, x_n = y; t_0, t_1, \ldots, t_{n-1})$$

from x to y. Since $d(x,z) < \eta \leq \delta$, we have

$$d(\phi^T(x), \phi^T(z)) < \epsilon.$$

So,

$$(z, \phi^T(x), x_1, \ldots, x_n = y; T, t_0 - T, t_1, \ldots, t_{n-1})$$

is an (ϵ, T)-chain from z to y.

To obtain an (ϵ, T)-chain from y to z, notice that there exists an $(\epsilon/2, T)$-chain

$$(y = x_0, \ldots, x_{n-1}, x_n = x; t_0, \ldots, t_{n-1})$$

from y to x. By the triangle inequality,

$$d(\phi^{t_{n-1}}(x_{n-1}), z) \leq d(\phi^{t_{n-1}}(x_{n-1}), x) + d(x,z) < \frac{\epsilon}{2} + \frac{\epsilon}{2} = \epsilon.$$

Thus,

$$(y = x_0, \ldots, x_{n-1}, z; t_0, \ldots, t_{n-1})$$

is an (ϵ, T)-chain from y to z.

Concatenating the (ϵ, T)-chain from z to y with the (ϵ, T)-chain from y to z we obtain an (ϵ, T)-chain from z to itself. Hence, z is chain recurrent and is chain equivalent to y. Thus, $z \in C$. Therefore, each chain component of ϕ^t is closed. □

It is worth noticing that the fact that chain components are closed does not imply that the chain recurrent set is closed, since a flow may have uncountably many chain components. In fact, we shall encounter a flow with uncountably many chain components in Example 3.3.14.

We now prove that a chain component is a maximal chain transitive set.

Proposition 3.3.10. (i) *Every chain component of a flow on a metric space is chain transitive with respect to the flow.*

(ii) *Every chain transitive set with respect to a flow on a metric space is a subset of a unique chain component of the flow.*

(iii) *If A and B are chain transitive with respect to a flow on a metric space, $A \subseteq B$ and C is the unique chain component containing A, then $B \subseteq C$.*

Proof. (i) A chain component is nonempty since it is an equivalence class. A chain component is closed by Proposition 3.3.9 and is invariant by Proposition 3.3.7. Furthermore, any two points in a chain component are chain equivalent by definition. Therefore, a chain component is a chain transitive set.

(ii) By interchanging the roles of x and y in Definition 3.3.1, any two points in a chain transitive set are chain equivalent. Thus, each chain transitive set lies in a chain component. Since chain components are equivalence classes, they are disjoint. Therefore, this chain component is unique.

(iii) By part (ii) the set B lies in a unique chain component of the flow. Since C is the unique chain component containing A and $A \subseteq B$ the set C is the unique chain component containing B. In particular, $B \subseteq C$. □

To demonstrate that the chain components of a flow on a compact metric space are exactly the connected components of the chain recurrent set we begin by showing that a chain component is a connected set.

Proposition 3.3.11. *Every chain component of a flow on a compact metric space is connected.*

Proof. Let C be a chain component of a flow ϕ^t on a compact metric space (X, d). By means of contradiction, assume that C is not connected. That is, there exist disjoint nonempty open subsets U and V of $\mathcal{R}(\phi^t)$ such that $C = U \cup V$.

Let $x \in U$ and $z \in V$. Let $\{\epsilon_k\}$ be a sequence of positive real numbers such that $\epsilon_k \to 0$ as $k \to \infty$. By Theorem 2.7.18 (iv), for each positive integer k there exists an ϵ_k-chain

$$x = y_{k,0}, \ldots, y_{k,n_k} = z$$

from x to z with respect to ϕ^1. Since $x \in U$ and $z \notin U$, for each positive integer k there is a first point in the chain $x = y_{k,0}, \ldots, y_{k,n_k} = z$ which is not an element of U. That is, for each positive integer k there exists an integer j_k such that $y_{k,j_k+1} \notin U$ and $y_{k,i} \in U$ for $i = 1, 2, 3, \ldots, j_k$. Define $x_k = y_{k,j_k}$ and $z_k = y_{k,j_k+1}$. Since X is compact, the sequences $\{x_k\}$ and $\{z_k\}$ have convergent subsequences (which, by reindexing, we also label

$\{x_k\}$ and $\{z_k\}$). Let $\bar{x} = \lim_{k\to\infty} x_k$ and $\bar{z} = \lim_{k\to\infty} z_k$. Since V is open in $\mathcal{R}(\phi^t)$, the set $U = \mathcal{R}(\phi^t) \setminus V$ is closed in $\mathcal{R}(\phi^t)$. The sequence x_k lies in U, so $\bar{x} \in U$. Similarly, $\bar{z} \in V$.

We will show that $\bar{z} \in U$. Let $\eta > 0$. Since X is compact and ϕ^1 is continuous, ϕ^1 is uniformly continuous. Thus, there exists $\delta > 0$ such that if $y_1, y_2 \in X$ and $d(y_1, y_2) < \delta$, then

$$d(\phi^1(y_1), \phi^1(y_2)) < \frac{\eta}{3}.$$

Let $\zeta = \min(\delta, \eta/3)$. There exists a positive integer N such that $\epsilon_N < \zeta$, $d(x_N, \bar{x}) < \zeta$ and $d(z_N, \bar{z}) < \zeta$. Because $\zeta \leq \delta$ we obtain

$$d(\phi^1(\bar{x}), \phi^1(x_N)) < \frac{\eta}{3}.$$

Since $d(\phi^1(x_N), z_N) < \epsilon_N < \zeta \leq \eta/3$ we have

$$d(\phi^1(\bar{x}), \bar{z}) < d(\phi^1(\bar{x}), \phi^1(x_N)) + d(\phi^1(x_N), z_N) + d(z_N, \bar{z}) < \eta$$

by the triangle inequality. According to Proposition 3.3.7, chain components are flow invariant. So, $\mathcal{O}(\bar{x}) \subseteq C = U \cup V$. Since $\mathcal{O}(\bar{x}) \cap U \neq \emptyset$ and $\mathcal{O}(\bar{x})$ is connected by Proposition 1.1.11, we have $\mathcal{O}(\bar{x}) \subseteq U$. In particular, $\phi^1(\bar{x}) \in U$. Hence, $\bar{z}$ is a limit point of U. The closedness of U in $\mathcal{R}(\phi^t)$ implies that $\bar{z} \in U$.

Thus, $U \cap V \neq \emptyset$, contradicting the assumption that C is not connected. Therefore, each connected component of $\mathcal{R}(\phi^t)$ is connected. □

Now that we know each chain component is connected, we shall prove that the chain components are the connected components of the chain recurrent set. It is precisely this interplay between chain transitivity and connectedness which motivated Conley [**11**] to refer to the equivalence classes of the chain equivalence relation as chain components.

Theorem 3.3.12. *The chain components of a flow on a compact metric space are the connected components of the chain recurrent set of the flow.*

Proof. Let C be a chain component of a flow ϕ^t on a compact metric space (X, d). By Proposition 3.3.11, C is connected. Thus, there is a unique connected component A of $\mathcal{R}(\phi^t)$ containing C. It suffices to show that $A \subseteq C$.

Let $y \in A$. We will prove that y is chain equivalent to each point in C. Let $x \in C$, $\epsilon > 0$ and $T > 0$. We shall produce (ϵ, T)-chains from x to y and from y to x. The set $\mathcal{R}(\phi^t)$ is closed by Proposition 2.7.10. Thus, A is closed because it is a connected component of a closed set. Since A is closed and X is compact, A is compact. Hence, there exist $z_1, \ldots, z_k \in A$

such that

$$A \subseteq \bigcup_{i=1}^{k} B(z_i, \epsilon/4).$$

Since A is connected, by reindexing if necessary, there is a finite collection

$$\{B(z_1, \epsilon/3), \ldots, B(z_n, \epsilon/3)\}$$

of balls such that $x \in B(z_1, \epsilon/3)$, $y \in B(z_n, \epsilon/3)$ and

$$B(z_i, \epsilon/3) \cap B(z_{i+1}, \epsilon/3) \neq \emptyset$$

for $i = 1, \ldots, n-1$.

The point x is chain recurrent; so, there is an $(\epsilon/3, T)$-chain

$$(x = x_0, \ldots, x_{n-1}, x_n = x; t_0, t_1, \ldots, t_{n-1})$$

from x to itself. By the triangle inequality,

$$\begin{aligned} d(\phi^{t_{n-1}}(x_{n-1}), z_1) &\leq d(\phi^{t_{n-1}}(x_{n-1}), x) + d(x, z_1) \\ &< \frac{\epsilon}{3} + \frac{\epsilon}{3} \\ &< \epsilon. \end{aligned}$$

Thus,

$$(x = x_0, \ldots, x_{n-1}, z_1; t_0, \ldots, t_{n-1})$$

is an (ϵ, T)-chain from x to z_1.

For each $i = 1, \ldots, n-1$ the point z_i is chain recurrent, so there is an $(\epsilon/3, T)$-chain

$$(z_i = w_0, \ldots, w_{n-1}, w_n = z_i; q_0, q_1, \ldots, q_{n-1})$$

from z_i to itself. By the triangle inequality,

$$\begin{aligned} d(\phi^{q_{n-1}}(w_{n-1}), z_{i+1}) &\leq d(\phi^{q_{n-1}}(w_{n-1}), z_i) + d(z_i, z_{i+1}) \\ &< \frac{\epsilon}{3} + \frac{2\epsilon}{3} \\ &= \epsilon. \end{aligned}$$

Thus,

$$(z_i = w_0, \ldots, w_{n-1}, z_{i+1}; q_0, \ldots, q_{n-1})$$

is an (ϵ, T)-chain from z_i to z_{i+1}. By concatenating these chains for $i = 1, \ldots, n-1$ we obtain an (ϵ, T)-chain from z_1 to z_n.

The point z_n is chain recurrent; so, there is an $(\epsilon/3, T)$-chain

$$(z_n = y_0, \ldots, y_{n-1}, y_n = z_n; s_0, s_1, \ldots, s_{n-1})$$

from z_n to itself. By the triangle inequality,

$$\begin{aligned} d(\phi^{s_{n-1}}(y_{n-1}), y) &\leq d(\phi^{s_{n-1}}(y_{n-1}), z_n) + d(z_n, y) \\ &< \frac{\epsilon}{3} + \frac{\epsilon}{3} \\ &< \epsilon. \end{aligned}$$

Thus,

$$(z_n = y_0, \ldots, y_{n-1}, y; s_0, \ldots, s_{n-1})$$

is an (ϵ, T)-chain from z_n to y.

Concatenate the chains from x to z_1, from z_1 to z_n, and from z_n to y to form an (ϵ, T)-chain from x to y. Similarly, there is an (ϵ, T)-chain from y to x. Thus, $y \in C$. So, $A \subseteq C$. Therefore, the chain components of ϕ^t are the connected components of $\mathcal{R}(\phi^t)$. □

Example 3.3.13. Let ψ^t be the arrested logistic rotation-dilation. Example 2.7.9 showed that

$$\mathcal{R}(\psi^t) = \partial\mathbb{D} \cup \{0\}.$$

The connected components of $\mathcal{R}(\psi^t)$ are $\partial\mathbb{D}$ and $\{0\}$. By Theorem 3.3.12 the chain components of ψ^t are $\partial\mathbb{D}$ and $\{0\}$ as well.

As the following example demonstrates, a flow may have uncountably many chain components.

Example 3.3.14. Let $C \subset [0,1]$ be the Cantor ternary set. Let $f : [0,1] \to \mathbb{R}$ be a smooth function such that $f(x) = 0$ for all $x \in C$ and $f(x) < 0$ for all $x \in [0,1] \setminus C$. Consider the differential equation

$$\dot{x} = f(x)$$

on $[0,1]$. If ϕ^t is the flow associated to this equation, then

$$\mathcal{R}(\phi^t) = C$$

by Example 2.7.8. Since the Cantor ternary set is totally disconnected, its connected components are precisely its singleton subsets. Therefore, because the Cantor ternary set is uncountable, the flow ϕ^t has uncountably many chain components.

3.4. Attracting Sets and Repelling Sets

The goal of this section is to define attracting sets and repelling sets and to describe chain recurrence in terms of these sets. The attracting sets and repelling sets of a flow arise from trapping regions. To motivate the definition of a trapping region consider the flow of a linear differential equation such that the real part of each eigenvalue of the coefficient matrix is negative.

Example 3.4.1. Let A be an $n \times n$ matrix with real entries such that the real part of each eigenvalue of A is negative. Consider the linear differential equation

$$\dot{x} = Ax$$

on $\mathbb{R}^n$. By Example 1.1.6 the flow ϕ^t of this equation is

$$\phi^t(x) = e^{At}x.$$

Since the real part of each eigenvalue of A is negative there exists a norm $\|\cdot\|_*$ on $\mathbb{R}^n$ equivalent to the standard norm on $\mathbb{R}^n$ and a positive real number a such that

$$\|e^{At}x\|_* \leq e^{-at}\|x\|_*$$

for all nonnegative real numbers t and $x \in \mathbb{R}^n$. See Robinson [**37**] for a proof of the existence of this adapted norm. Because zero is not an eigenvalue of A, the kernel of A is $\{0\}$. Thus, the origin is the only fixed point of ϕ^t by Example 2.1.4. Denote the open ball of radius $r > 0$ centered at the origin with respect to $\|\cdot\|_*$ by $B(0,r)$. If $x \in \overline{B(0,r)}$, then

$$\|\phi^t(x)\|_* = \|e^{At}x\|_* \leq e^{-at}\|x\|_* \leq r$$

for all nonnegative real numbers t. Thus, $\overline{B(0,r)}$ is positively invariant with respect to ϕ^t. Moreover, $\phi^1(\overline{B(0,r)}) \subseteq B(0,r)$.

In the spirit of Example 3.4.1 a set is a trapping region for a flow if it is positively invariant and there is a uniform positive time by which the orbits of all points in its closure enter its interior.

Definition 3.4.2. Let ϕ^t be a flow on a topological space X. A set $U \subseteq X$ is a **trapping region** for ϕ^t if U is positively invariant with respect to ϕ^t and there exists a positive real number T such that

$$\phi^T(\mathrm{cl}(U)) \subseteq \mathrm{int}(U).$$

See Figure 3.2.

Example 3.4.3. If ϕ^t is a flow on a topological space X, then $\emptyset$ and X are trapping regions for ϕ^t.

The interior of a nonempty trapping region is nonempty.

Example 3.4.4. Let A be an $n \times n$ matrix with real entries such that the real part of each eigenvalue of A is negative. Example 3.4.1 shows that balls centered at the origin with respect to the adapted norm are trapping regions for the flow on $\mathbb{R}^n$ corresponding to the linear differential equation $\dot{x} = Ax$.

Example 3.4.5. Let a and b be real numbers with $a < b$. Consider a smooth function $f : [a,b] \to \mathbb{R}$ such that $f(a) = f(b) = 0$ and $f(x) < 0$ for all $x \in (a,b)$. Let ϕ^t denote the decreasing segment flow generated by f on

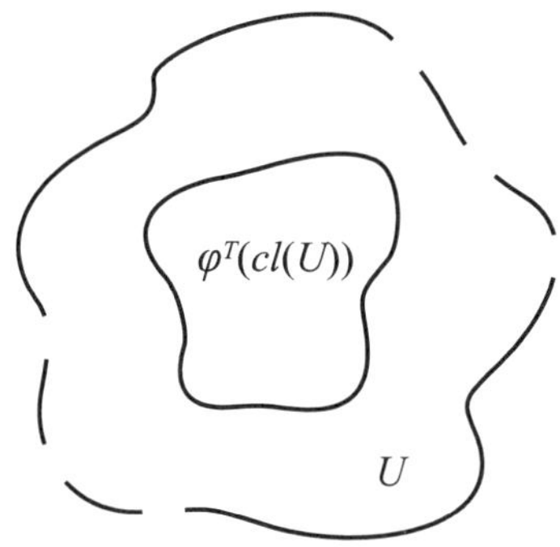

Figure 3.2. A trapping region

$[a, b]$. Recall that the decreasing segment flow corresponds to the differential equation

$$\dot{x} = f(x)$$

on $[a, b]$. We will show that the nonempty closed trapping regions for ϕ^t are exactly the intervals of the form $[a, c]$ such that $a < c \leq b$.

The interval $[a, b]$ is a closed trapping region for ϕ^t by Example 3.4.3. Let $c \in (a, b)$. To show that $U = [a, c]$ is a trapping region for ϕ^t, we make three preliminary observations. First, the negativity of f on (a, b) implies that for each $x \in (a, b)$ the function $\phi^t(x)$ is strictly decreasing with respect to t. In particular, $\phi^1(c) < c$. Second, $\text{Fix}(\phi^t) = \{a, b\}$ by Example 2.1.3. Third, since $\phi^1 : [a, b] \to [a, b]$ is a homeomorphism and $\text{Fix}(\phi^t) = \{a, b\}$, the map ϕ^1 preserves orientation. That is, if $a \leq x < y \leq b$, then $a \leq \phi^1(x) < \phi^1(y) \leq b$.

Now, if $x \in [a, c]$ and $t \geq 0$, then since $a \in \text{Fix}(\phi^t)$ and $\phi^t(x)$ is decreasing with respect to t we obtain

$$a \leq \phi^t(x) \leq x \leq c.$$

Thus, $\phi^t(x) \in U$, so that U is positively invariant.

Next, we claim that $\phi^1(\text{cl}(U)) \subseteq \text{int}(U)$. To see this, notice that if $x \in \text{cl}(U) = [a, c]$, then the fact that ϕ^1 preserves orientation and is strictly decreasing on (a, b) implies that

$$a \leq \phi^1(x) \leq \phi^1(c) < c,$$

so that $\phi^1(x) \in [a, c) = \text{int}(U)$. Thus, $\phi^1(\text{cl}(U)) \subseteq \text{int}(U)$. Therefore, $U = [a, c]$ is a trapping region for ϕ^t. See Figure 3.3.

Let U be a nonempty closed trapping region for ϕ^t. The set U is nonempty and bounded above by b. By the completeness of $[a, b]$, the set U has a supremum c. Since U is positively invariant and closed, $\text{cl}(\mathcal{O}^+(c)) =$

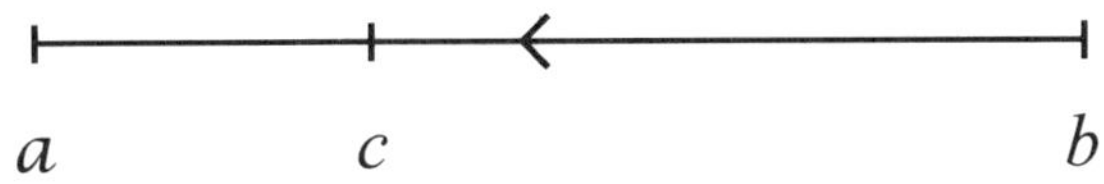

Figure 3.3. A trapping region for the segment flow

$[a, c] \subseteq U$. The singleton $\{a\}$ is not a trapping region because a nonempty trapping region must have a nonempty interior. Thus, $c \neq a$. Therefore, every nonempty closed trapping region U for ϕ^t is an interval of the form $[a, c]$ such that $a < c \leq b$.

Example 3.4.6. Consider the flow ϕ^t corresponding to the system of differential equations

$$\begin{aligned} \dot{x}_1 &= 1, \\ \dot{x}_2 &= \sin(2\pi x_2) \end{aligned}$$

on the torus $\mathbb{T}^2 = \mathbb{R}^2/\mathbb{Z}^2$. We shall refer to this flow as the **twist flow**. Let $\delta \in (0, 1/2)$. We claim that

$$U = \left\{(x, y) \in \mathbb{T}^2 \mid \tfrac{1}{2} - \delta < y < \tfrac{1}{2} + \delta\right\}$$

is a trapping region for ϕ^t.

Let $(x, y) \in U$. If $y = 1/2$, then $\dot{x}_2 = 0$, so that $x_2 = 1/2$. If $1/2 < y < 1/2 + \delta$, then $\dot{x}_2 < 0$, so that x_2 is decreasing to $1/2$. If $1/2 - \delta < y < 1/2$, then $\dot{x}_2 > 0$, so that x_2 is increasing to $1/2$. Thus, U is positively invariant.

Since ∂U is compact and $\dot{x}_2 \neq 0$ for initial conditions on ∂U, there is a positive real number T such that $\phi^T(\mathrm{cl}(U)) \subset \mathrm{int}(U)$. Therefore, U is a trapping region for ϕ^t. See Figure 3.4.

The following example will be helpful in the proof of Theorem 3.4.27.

Example 3.4.7. Let ϕ^t be a flow on a compact metric space (X, d). Let $y \in X$, $\epsilon > 0$ and $T > 0$. Define

$$U = \{x \in X \mid \forall s \in [0, T] \text{ there is an } (\epsilon, T)\text{-chain from } y \text{ to } \phi^{-s}(x)\}.$$

We claim that U is a trapping region for ϕ^t.

To show that U is positively invariant with respect to ϕ^t, let $\tau > 0$ and let $u \in U$. It suffices to show that for every $s \in [0, T]$ there exists an (ϵ, T)-chain from y to $\phi^\tau(\phi^{-s}(u)) = \phi^{\tau-s}(u)$. If $s - \tau \geq 0$, then $s - \tau \in [0, T]$. By the definition of U there exists an (ϵ, T)-chain from y to $\phi^{\tau-s}(u)$. If $\tau - s > 0$, then there is an (ϵ, T)-chain

$$(y = x_0, x_1, \ldots, x_n = \phi^{-T}(u); t_0, t_1, \ldots, t_{n-1})$$

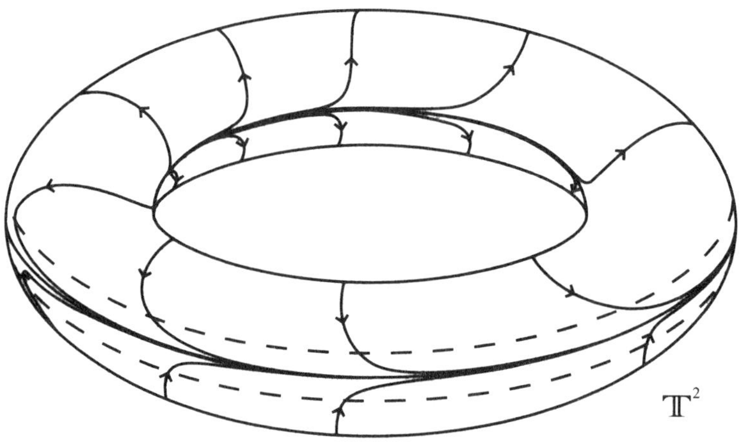

Figure 3.4. A trapping region for the twist flow

from y to $\phi^{-T}(u)$, in which case

$$(y = x_0, x_1, \ldots, x_n = \phi^{-T}(u), \phi^{\tau-s}(u); t_0, t_1, \ldots, t_{n-1}, T + \tau - s)$$

is an (ϵ, T)-chain from y to $\phi^{\tau-s}(u)$. Thus, U is positively invariant with respect to ϕ^t.

Next we shall show that

$$\phi^{2T}(\mathrm{cl}(U)) \subseteq \mathrm{int}(U).$$

Let $u \in \mathrm{cl}(U)$, and let $z = \phi^{2T}(u)$. By the continuity of ϕ^t and compactness of $[-T, 0] \times X$ and $[T, 2T] \times X$, there exists $\delta > 0$ such that

(i) if $d(z, w) < \delta$, then

$$d(\phi^\tau(z), \phi^\tau(w)) < \frac{\epsilon}{2}$$

for all $\tau \in [-T, 0]$, and

(ii) if $d(u, v) < \delta$, then

$$d(\phi^\tau(u), \phi^\tau(v)) < \frac{\epsilon}{2}$$

for all $\tau \in [T, 2T]$.

Let $w \in B(u, \delta)$. We will show that $w \in U$, so that $z \in \mathrm{int}(U)$. Let $s \in [0, T]$. We shall produce an (ϵ, T)-chain from y to $\phi^{-s}(w)$. Since $u \in \mathrm{cl}(U)$ there exists $v \in U$ with $d(u, v) < \delta$. Since $v \in U$, there is an (ϵ, T)-chain

$$(y = x_0, x_1, \ldots, x_n = v; t_0, t_1, \ldots, t_{n-1})$$

from y to v. Notice that $2T - s \in [T, 2T]$. According to the choice of $\delta > 0$,

$$\begin{aligned} d(\phi^{2T-s}(v), \phi^{-s}(w)) &\leq d(\phi^{2T-s}(v), \phi^{2T-s}(u)) + d(\phi^{2T-s}(u), \phi^{-s}(w)) \\ &\leq d(\phi^{2T-s}(v), \phi^{2T-s}(u)) + d(\phi^{-s}(z), \phi^{-s}(w)) \\ &< \frac{\epsilon}{2} + \frac{\epsilon}{2} = \epsilon. \end{aligned}$$

Thus,

$$(y = x_0, x_1, \ldots, x_n = v, \phi^{-s}(w); t_0, t_1, \ldots, t_{n-1}, 2T - s)$$

is an (ϵ, T)-chain from y to $\phi^{-s}(w)$. So, $w \in U$ and $z \in \mathrm{int}(U)$. Consequently,

$$\phi^{2T}(\mathrm{cl}(U)) \subseteq \mathrm{int}(U).$$

Therefore, U is a trapping region for ϕ^t.

A trapping region determines two sets called the attracting set and the repelling set, respectively, associated to the trapping region.

Definition 3.4.8. Let ϕ^t be a flow on a topological space X. A set A is an **attracting set** for ϕ^t if there exists a trapping region U for ϕ^t such that

$$A = \bigcap_{t \geq 0} \phi^t(U).$$

A set A^* is a **repelling set** for ϕ^t if there exists a trapping region U for ϕ^t such that

$$A^* = \bigcap_{t \leq 0} \phi^t(X \setminus U).$$

When we wish to emphasize the dependence of an attracting set A or a repelling set A^* on the trapping region U from which it arises, we denote it by A_U or A_U^*, respectively.

Conley [**11**] introduced attracting sets and repelling sets, referring to attracting sets as attractors and to repelling sets as repellers. We reserve the terms attractor and repeller for specific types of attracting sets and repelling sets. See Definition 3.4.22.

Definition 3.4.9. Let ϕ^t be a flow on a topological space X. A pair (A, A^*) of subsets of X is an **attracting-repelling pair** for ϕ^t if there exists a trapping region U for ϕ^t such that

$$A = \bigcap_{t \geq 0} \phi^t(U),$$

and

$$A^* = \bigcap_{t \leq 0} \phi^t(X \setminus U).$$

Example 3.4.10. For a flow on a topological space X the pairs $(\emptyset, X)$ and $(X, \emptyset)$ are attracting-repelling pairs.

Example 3.4.11. Let A be an $n \times n$ matrix with real entries such that the real part of each eigenvalue of A is negative. Let ϕ^t be the flow of the linear differential equation

$$\dot{x} = Ax$$

on $\mathbb{R}^n$. By Example 3.4.1, a closed ball U centered at the origin with respect to the adapted norm is a trapping region for ϕ^t. The attracting set corresponding to U is

$$\bigcap_{t \geq 0} \phi^t(U) = \{0\}.$$

The associated repelling set is the empty set. Therefore, $(\{0\}, \emptyset)$ is an attracting-repelling pair for ϕ^t.

Example 3.4.12. Let ϕ^t be the decreasing segment flow on $[a,b]$. By Example 3.4.5 the closed trapping regions for ϕ^t are exactly the sets $\emptyset$ and $[a,c]$ for $a < c \leq b$. By Example 3.4.10 the attracting-repelling pairs arising from $U = \emptyset$ and $U = [a,b]$ are $(\emptyset, [a,b])$ and $([a,b], \emptyset)$, respectively.

If $U = [a,c)$ for some $c \in (a,b)$ and

$$A = \bigcap_{t \geq 0} \phi^t(\mathrm{cl}(U)),$$

then $a \in A$ and $b \in [a,b] \setminus A$. Since $\lim_{t \to \infty} \phi^t(c) = a$, if $x \in (a,b)$, then $x \notin \phi^t(\mathrm{cl}(U))$ for some $t > 0$. Therefore, $A = \{a\}$. Similarly, $A^* = \{b\}$.

Thus, the attracting-repelling pairs corresponding to the closed trapping regions of ϕ^t are $([a,b], \emptyset)$, $(\{a\}, \{b\})$ and $(\emptyset, [a,b])$.

Example 3.4.13. Consider the twist flow ϕ^t on the torus $\mathbb{T}^2 = \mathbb{R}^2/\mathbb{Z}^2$. Example 3.4.6 showed that for each $\delta \in (0, 1/2)$ the set

$$U = \left\{(x,y) \in \mathbb{T}^2 \mid \tfrac{1}{2} - \delta < y < \tfrac{1}{2} + \delta\right\}$$

is a trapping region for ϕ^t. The corresponding attracting-repelling pair is

$$\left(\left\{(x,y) \in \mathbb{T}^2 \mid y = \tfrac{1}{2}\right\}, \left\{(x,y) \in \mathbb{T}^2 \mid y = 0\right\}\right).$$

Proposition 3.4.14. *Let ϕ^t be a flow on a topological space X. If U is a trapping region for ϕ^t associated to the attracting-repelling pair (A, A^*) for ϕ^t, then $X \setminus U$ is a trapping region for ϕ^{-t} and (A^*, A) is the attracting-repelling pair for ϕ^{-t} associated to $X \setminus U$.*

Proof. Since U is a trapping region for ϕ^t the set U is positively invariant with respect to ϕ^t. By Proposition 1.4.4 (i) the set $X \setminus U$ is negatively invariant with respect to ϕ^t. That is, $X \setminus U$ is positively invariant with respect to ϕ^{-t}.

Since U is a trapping region for ϕ^t there exists a positive real number T such that

$$\phi^T(\mathrm{cl}((U)) \subseteq \mathrm{int}(U).$$

Thus,

$$X \setminus \mathrm{int}(U) \subseteq X \setminus \phi^T(\mathrm{cl}(U)) = \mathrm{int}(X \setminus \phi^T(U)) = \phi^T(\mathrm{int}(X \setminus U)).$$

Consequently,

$$\phi^{-T}(\mathrm{cl}(X \setminus U)) = \phi^{-T}(X \setminus \mathrm{int}(U)) \subseteq \mathrm{int}(X \setminus U).$$

Therefore, $X \setminus U$ is a trapping region for ϕ^{-t}.

Since (A, A^*) is the attracting-repelling pair associated to ϕ^t,

$$A_{X\setminus U} = \bigcap_{t\geq 0} \phi^{-t}(X \setminus U) = \bigcap_{t\leq 0} \phi^t(X \setminus U) = A^*,$$

and

$$A^*_{X\setminus U} = \bigcap_{t\leq 0} \phi^{-t}(X \setminus (X \setminus U)) = \bigcap_{t\geq 0} \phi^t(U) = A.$$

Therefore, (A^*, A) is the attracting-repelling pair for ϕ^{-t} associated to $X \setminus U$. □

Proposition 3.4.15 and Corollary 3.4.16 guarantee that there is no loss of generality in considering only open or closed trapping regions when concerned with attracting sets or repelling sets.

Proposition 3.4.15. *Let ϕ^t be a flow on a topological space X. If A is the attracting set for ϕ^t associated to the trapping region U, then*

$$A = \bigcap_{t\geq 0} \phi^t(cl(U)) = \bigcap_{t\geq 0} \phi^t(int(U)).$$

If A^* is the repelling set for ϕ^t associated to the trapping region U, then

$$A^* = \bigcap_{t\leq 0} \phi^t(X \setminus int(U)) = \bigcap_{t\leq 0} \phi^t(X \setminus cl(U)).$$

Proof. Let A be the attracting set associated to a trapping region U for ϕ^t. Since U is a trapping region there exists $T > 0$ such that $\phi^T(\mathrm{cl}(U)) \subseteq \mathrm{int}(U)$, and

$$A = \bigcap_{t\geq 0} \phi^t(U) \subseteq \bigcap_{t\geq 0} \phi^t(\mathrm{cl}(U)) \subseteq \bigcap_{t\geq 0} \phi^{t+T}(\mathrm{cl}(U)) \subseteq \bigcap_{t\geq 0} \phi^t(\mathrm{int}(U)) \subseteq A.$$

Therefore,

$$A = \bigcap_{t\geq 0} \phi^t(\mathrm{cl}(U)) = \bigcap_{t\geq 0} \phi^t(\mathrm{int}(U)).$$

Similarly,

$$A^* = \bigcap_{t \leq 0} \phi^t(X \setminus \mathrm{int}(U)) = \bigcap_{t \leq 0} \phi^t(X \setminus \mathrm{cl}(U)). \qquad \square$$

Corollary 3.4.16. *Let ϕ^t be a flow on a topological space X. If A is an attracting set for ϕ^t, then there exists an open trapping region V for ϕ^t and there exists a closed trapping region W for ϕ^t such that*

$$A = \bigcap_{t \geq 0} \phi^t(W) = \bigcap_{t \geq 0} \phi^t(V).$$

If A^* is a repelling set for ϕ^t, then there exists an open trapping region V for ϕ^t and there exists a closed trapping region W for ϕ^t such that

$$A^* = \bigcap_{t \leq 0} \phi^t(X \setminus V) = \bigcap_{t \leq 0} \phi^t(X \setminus W).$$

Proof. Since A is an attracting set there exists a trapping region U for ϕ^t such that

$$A = \bigcap_{t \geq 0} \phi^t(U).$$

Let $V = \mathrm{int}(U)$ and $W = \mathrm{cl}(U)$. The sets V and W are open and closed, respectively. By Proposition 3.4.15,

$$A = \bigcap_{t \geq 0} \phi^t(W) = \bigcap_{t \geq 0} \phi^t(V).$$

Similarly,

$$A^* = \bigcap_{t \leq 0} \phi^t(X \setminus V) = \bigcap_{t \leq 0} \phi^t(X \setminus W). \qquad \square$$

Example 3.4.17. Let ϕ^t be the decreasing segment flow on $[a, b]$. By Example 3.4.12 the attracting-repelling pairs corresponding to the closed trapping regions for ϕ^t are $([a, b], \emptyset)$, $(\{a\}, \{b\})$, and $(\emptyset, [a, b])$. By Corollary 3.4.16 these are all of the attracting-repelling pairs for ϕ^t.

A further corollary of Proposition 3.4.15 shows that it suffices to consider large times when obtaining an attracting set from a trapping region.

Corollary 3.4.18. *Let ϕ^t be a flow on a topological space. If A is the attracting set for ϕ^t associated to a trapping region U for ϕ^t, then*

$$A = \bigcap_{t \geq T} \phi^t(cl(U))$$

for every positive real number T such that $\phi^T(cl(U)) \subseteq int(U)$.

Proof. Let U be a trapping region for ϕ^t, and let T be a positive real number such that $\phi^T(\mathrm{cl}(U)) \subseteq \mathrm{int}(U)$. If A is the attracting set associated to U, then

$$\begin{aligned} A &= \bigcap_{t\geq 0} \phi^t(\mathrm{cl}(U)) \\ &\subseteq \bigcap_{t\geq T} \phi^t(\mathrm{cl}(U)) \\ &= \bigcap_{t\geq 0} \phi^{t+T}(\mathrm{cl}(U)) \\ &= \bigcap_{t\geq 0} \phi^t(\phi^T(\mathrm{cl}(U)) \\ &\subseteq \bigcap_{t\geq 0} \phi^t(\mathrm{int}(U)) \\ &= A. \end{aligned}$$

Therefore,

$$A = \bigcap_{t\geq T} \phi^t(\mathrm{cl}(U)). \qquad \square$$

Attracting sets and repelling sets are closed and flow invariant.

Proposition 3.4.19. *Every attracting set for a flow is closed and invariant with respect to the flow. Every repelling set for a flow is closed and invariant with respect to the flow.*

Proof. Let A be an attracting set for a flow ϕ^t. By Corollary 3.4.16 there exists a closed trapping region U such that

$$A = \bigcap_{t\geq 0} \phi^t(U).$$

Since ϕ^t is a homeomorphism for each $t \in \mathbb{R}$, the set $\phi^t(U)$ is closed for each $t \in \mathbb{R}$. So, A is an intersection of closed sets. Therefore, A is closed. Similarly, A^* is closed.

Let $x \in A$, and let $s \in \mathbb{R}$. To prove that A is invariant we must show that $\phi^s(x) \in \phi^t(U)$ for all $t \geq 0$. Let $t \geq 0$. If $s < 0$, then, since $x \in A$ and $t - s \geq 0$, we have $x \in \phi^{t-s}(U)$. Thus,

$$\phi^s(x) \in \phi^s(\phi^{t-s}(U)) = \phi^t(U).$$

If $s \geq 0$, then

$$\phi^s(x) \in \phi^s(\phi^t(U))$$

since $x \in A$, so that $x \in \phi^t(U)$. Because U is positively invariant and $s \geq 0$ we obtain

$$\phi^s(\phi^t(U)) = \phi^t(\phi^s(U)) \subseteq \phi^t(U).$$

Consequently,

$$\phi^s(x) \in \phi^t(U).$$

Therefore, A is invariant. Similarly, A^* is invariant. □

Attracting sets and repelling sets capture the eventual behavior of every orbit of a flow.

Proposition 3.4.20. *Let ϕ^t be a flow on a Hausdorff topological space X. If (A, A^*) is an attracting-repelling pair for ϕ^t and $x \in X \setminus A^*$, then $\omega(x) \subseteq A$. If $x \in X \setminus A$, then $\alpha(x) \subseteq A^*$.*

Proof. Since A is an attracting set there exists a trapping region U for ϕ^t such that

$$A = \bigcap_{t \geq 0} \phi^t(U).$$

Let $x \in X \setminus A^*$, and let $y \in \omega(x)$. We must show that $y \in A$.

Let $t \geq 0$. Since $x \in X \setminus A^*$, there exists $s \leq 0$ with $x \notin \phi^s(X \setminus U)$. Consequently, $\phi^{-s}(x) \in U$, and $\phi^t(\phi^{-s}(x)) \in \phi^t(U)$. Because $y \in \omega(x)$, there exists a sequence t_n of real numbers such that $t_n \to \infty$ as $n \to \infty$ and

$$y = \lim_{n \to \infty} \phi^{t_n}(x).$$

So, there exists a positive integer N such that if $n > N$, then $t_n \geq t - s$.

The set U is a trapping region; hence, U is positively invariant. Since $t_n \geq t - s$ for all positive integers $n > N$ we obtain

$$\phi^{t_n}(x) \in \phi^t(U)$$

for all positive integers $n > N$. Thus, $y \in \mathrm{cl}(\phi^t(U)) = \phi^t(\mathrm{cl}(U))$. The number $t \geq 0$ was arbitrary, so $y \in A$ by Proposition 3.4.15. Therefore, $\omega(x) \subseteq A$. A similar argument proves that if $x \in X \setminus A$, then $\alpha(x) \subseteq A^*$. □

Example 3.4.21. Consider the twist flow ϕ^t on the torus $\mathbb{T}^2 = \mathbb{R}^2/\mathbb{Z}^2$. Example 3.4.13 showed that one attracting-repelling pair for ϕ^t is

$$\left(\left\{(x, y) \in \mathbb{T}^2 \,|\, y = \tfrac{1}{2}\right\}, \left\{(x, y) \in \mathbb{T}^2 \,|\, y = 0\right\}\right).$$

We shall demonstrate that every nonempty trapping region V for ϕ^t corresponds to this same attracting-repelling pair.

Let V be a nonempty trapping region for ϕ^t. There exists a positive real number T such that $\phi^T(\mathrm{cl}(V)) \subseteq \mathrm{int}(V)$. The interior of V is nonempty. So, there exists $(x, y) \in V$ with $y \notin \{0, 1/2\}$. Thus, $(x, y) \in \mathbb{T}^2 \setminus A^*$. By Proposition 3.4.20 we know that $\alpha(x, y) \subseteq A^*$. So, $(x, 0) \in A^*$. By

Proposition 3.4.19 the set A^* is invariant. Therefore, $\{(x,y) \in \mathbb{T}^2 \mid y = 0\} \subseteq A^*$.

Because V is nonempty there exists $(x,y) \in \mathbb{T}^2 \setminus \phi^T(\mathrm{cl}(V))$. But, $(x,1/2) \in \omega(x,y) \subseteq A$ by Proposition 3.4.20. Hence, by Proposition 3.4.19 we have $\{(x,y) \in \mathbb{T}^2 \mid y = 1/2\} \subseteq A$.

If there exists $(x,y) \in A$ with $y \neq 1/2$, then by Proposition 3.4.19 both $\alpha(x,y)$ and $\omega(x,y)$ are subsets of A. In particular, $(x,0) \in A$, contradicting the fact that A and A^* are disjoint. Similarly, we obtain a contradiction if there is an $(x,y) \in A^*$ with $y \notin \{0,1\}$. Hence, the attracting-repelling pair corresponding to V must be

$$\left(\left\{(x,y) \in \mathbb{T}^2 \,|\, y = \tfrac{1}{2}\right\}, \left\{(x,y) \in \mathbb{T}^2 \,|\, y = 0\right\}\right).$$

We now consider attracting sets and repelling sets which are dynamically irreducible.

Definition 3.4.22. Let ϕ^t be a flow on a topological space. An **attractor** for ϕ^t is a nonempty chain transitive attracting set for ϕ^t. A **repeller** for ϕ^t is a nonempty chain transitive repelling set for ϕ^t.

There is considerable variation in the literature regarding the definitions of the terms attractor and repeller. Definition 3.4.22 is compatible with our development.

Example 3.4.23. Let A be an $n \times n$ matrix with real entries such that the real part of each eigenvalue of A is negative. Let ϕ^t be the flow of the linear differential equation

$$\dot{x} = Ax$$

on $\mathbb{R}^n$. By Example 3.4.11 the set $\{0\}$ is an attracting set for ϕ^t. The set $\{0\}$ is an attractor since a singleton fixed point is chain transitive.

Example 3.4.24. Let ϕ^t be the decreasing segment flow on $[a,b]$. By Example 3.4.17 the set $\{a\}$ is an attracting set for ϕ^t. Since a singleton fixed point is chain transitive, the set $\{a\}$ is an attractor for ϕ^t. Similarly, $\{b\}$ is a repeller for ϕ^t.

Example 3.4.25. Consider the twist flow ϕ^t on the torus $\mathbb{T}^2 = \mathbb{R}^2/\mathbb{Z}^2$. The twist flow corresponds to the system of differential equations

$$\begin{aligned} \dot{x}_1 &= 1, \\ \dot{x}_2 &= \sin(2\pi x_2). \end{aligned}$$

Example 3.4.13 showed that $\{(x,y) \in \mathbb{T}^2 \mid y = 1/2\}$ is an attracting set for ϕ^t. Since $\dot{x}_1 = 1$, this set is a periodic orbit. Thus, it is chain transitive. Therefore, it is an attractor for ϕ^t. Similarly, $\{(x,y) \in \mathbb{T}^2 \mid y = 0\}$ is a repeller for ϕ^t. See Figure 3.5.

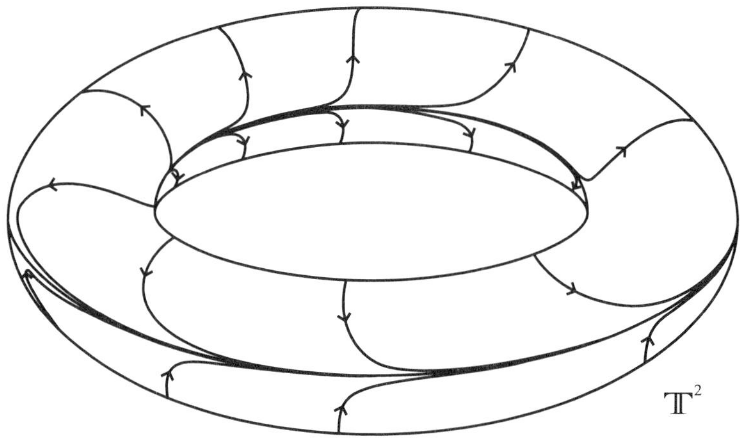

Figure 3.5. A phase portrait of the flow in Example 3.4.25

Next, we characterize the chain recurrent set of a flow on a compact metric space via the attracting-repelling pairs of the flow. Specifically, Conley [**11**] showed that if ϕ^t is a flow on a compact metric space, then

$$\mathcal{R}(\phi^t) = \bigcap \{A_U \cup A_U^* \mid U \text{ is a trapping region for } \phi^t\}.$$

Lemma 3.4.26. *If ϕ^t is a flow on a compact metric space (X, d) and U is a trapping region of ϕ^t, then*

$$\mathcal{R}(\phi^t) \cap U \subseteq A_U.$$

Proof. Let $x \in \mathcal{R}(\phi^t) \cap U$. We seek to prove that

$$x \in A_U = \bigcap_{t \geq 0} \phi^t(\text{cl}(U)).$$

Since U is a trapping region for ϕ^t, there exists $T > 0$ such that

$$\phi^T(\text{cl}(U)) \subseteq \text{int}(U).$$

By Corollary 3.4.18 it suffices to prove that $x \in \phi^\tau(\text{cl}(U))$ for every $\tau \geq T$.

Let $\tau \geq T$. Since $\phi^T(\text{cl}(U))$ is closed and X is compact, $\phi^T(\text{cl}(U))$ is compact. Thus, there exists $\eta > 0$ such that if $y \in \phi^T(\text{cl}(U))$, $z \in X$, and $d(y, z) < \eta$, then $z \in U$.

Let $\epsilon > 0$. Without loss of generality, we may assume that $\epsilon < \eta$. Since $x \in \mathcal{R}(\phi^t)$ there exists an (ϵ, τ)-chain

$$(x = x_0, x_1, \ldots, x_n = x; t_0, t_1, \ldots, t_{n-1})$$

from x to itself. Since U is positively invariant, $x_0 \in U$ and $t_0 \geq \tau \geq T$,

$$\phi^{t_0}(x_0) \in \phi^\tau(U) \subseteq \phi^T(U) \subseteq \phi^T(\text{cl}(U)).$$

Because x_0 and x_1 are in the (ϵ, τ)-chain we have

$$d(\phi^{t_0}(x_0), x_1) < \epsilon < \eta.$$

Thus, $x_1 \in U$. Similarly, $\phi^{t_1}(x_1) \in \phi^T(\mathrm{cl}(U))$, and $d(\phi^{t_1}(x_0), x_2) < \eta$. So, $x_2 \in U$. Continuing in this fashion we obtain $x_i \in U$ for $i = 0, 1, \ldots, n-1$.

Consequently,

$$\phi^{t_{n-1}}(x_{n-1}) \in \phi^\tau(\mathrm{cl}(U)),$$

and

$$d(\phi^{t_{n-1}}(x_{n-1}), x) < \epsilon.$$

Hence, x is a limit point of $\phi^\tau(\mathrm{cl}(U))$. Since $\phi^\tau(\mathrm{cl}(U))$ is closed, $x \in \phi^\tau(\mathrm{cl}(U))$. Thus, $x \in A_U$. Therefore,

$$\mathcal{R}(\phi^t) \cap U \subseteq A_U. \qquad \square$$

Theorem 3.4.27. *If ϕ^t is a flow on a compact metric space (X, d), then*

$$\mathcal{R}(\phi^t) = \bigcap \{A_U \cup A_U^* \mid U \text{ is a trapping region for } \phi^t\}.$$

Proof. Let

$$\mathcal{P} = \bigcap \{A_U \cup A_U^* \mid U \text{ is a trapping region for } \phi^t\}.$$

We begin by showing that $\mathcal{R}(\phi^t) \subseteq \mathcal{P}$.

Let $x \in \mathcal{R}(\phi^t)$, and let U be a trapping region for ϕ^t. Either $x \in U$ or $x \in X \setminus U$. If $x \in U$, then $x \in A_U$ by Lemma 3.4.26. If $x \in X \setminus U$, then $x \in A_U^*$ by Proposition 3.4.14. So, $\mathcal{R}(\phi^t) \subseteq A_U \cup A_U^*$ for every trapping region U for ϕ^t. Therefore, $\mathcal{R}(\phi^t) \subseteq \mathcal{P}$.

To prove the reverse inclusion, let $y \in X \setminus \mathcal{R}(\phi^t)$, so that there exist $\epsilon > 0$ and $T > 0$ such that there is no (ϵ, T)-chain from y to itself. Define

$$U = \{x \in X \mid \forall s \in [0, T] \text{ there is an } (\epsilon, T)\text{-chain from } y \text{ to } \phi^{-s}(x)\}.$$

By Example 3.4.7 the set U is a trapping region for ϕ^t. Since $y \notin U$ and $A_U \subseteq U$, we have $y \notin A_U$. For every $s \in [0, T]$,

$$(y, \phi^{2T-s}(y); 2T - s)$$

is an (ϵ, T)-chain from y to $\phi^{2T-s}(y)$. So, $\phi^{2T}(y) \in U$. That is, $y \in \phi^{-2T}(U)$. Thus, $y \notin \phi^{-2T}(X \setminus U)$, so that $y \notin A_U^*$. Hence, $y \notin \mathcal{P}$ because U is a trapping region for ϕ^t such that $y \notin A_U \cup A_U^*$. Therefore, $\mathcal{P} \subseteq \mathcal{R}(\phi^t)$. Consequently,

$$\mathcal{R}(\phi^t) = \mathcal{P} = \bigcap \{A_U \cup A_U^* \mid U \text{ is a trapping region for } \phi^t\}. \qquad \square$$

The following examples apply Theorem 3.4.27.

Example 3.4.28. Let ϕ^t be the decreasing segment flow on $[a, b]$. By Example 3.4.17 the attracting-repelling pairs for ϕ^t are $([a, b], \emptyset)$, $(\{a\}, \{b\})$ and $(\emptyset, [a, b])$. By Theorem 3.4.27,

$$\begin{aligned}\mathcal{R}(\phi^t) &= ([a, b] \cup \emptyset) \cap (\{a\} \cup \{b\}) \cap (\emptyset \cup [a, b]) \\ &= \{a, b\}.\end{aligned}$$

Recall that Example 2.7.7 also showed that $\mathcal{R}(\phi^t) = \{a, b\}$.

Example 3.4.29. Consider the twist flow ϕ^t on the torus $\mathbb{T}^2 = \mathbb{R}^2/\mathbb{Z}^2$. By Example 3.4.21 the only nontrivial attracting-repelling pair for ϕ^t is

$$\left(\left\{(x, y) \in \mathbb{T}^2 \,|\, y = \tfrac{1}{2}\right\}, \left\{(x, y) \in \mathbb{T}^2 \,|\, y = 0\right\}\right).$$

Therefore,

$$\mathcal{R}(\phi^t) = \left\{(x, y) \in \mathbb{T}^2 \mid y = \tfrac{1}{2}\right\} \cup \left\{(x, y) \in \mathbb{T}^2 \mid y = 0\right\}$$

by Theorem 3.4.27.

In addition to characterizing the chain recurrent set of a flow, attracting-repelling pairs allow us to distinguish between a flow's chain components. Specifically, if two points are chain equivalent, then for each attracting-repelling pair the two points are either both elements of the attracting set or both elements of the repelling set of the attracting-repelling pair. Consequently, for each chain component and each attracting-repelling pair either the chain component lies entirely within the attracting set or entirely within the repelling set of the attracting-repelling pair.

Proposition 3.4.30. *Let ϕ^t be a flow on a compact metric space (X, d). The points $x, y \in X$ are chain equivalent with respect to ϕ^t if and only if $x, y \in A$ or $x, y \in A^*$ for each attracting-repelling pair (A, A^*) for ϕ^t.*

Proof. Let $x, y \in X$ be chain equivalent with respect to ϕ^t, and let (A, A^*) be an attracting-repelling pair for ϕ^t. By Corollary 3.4.16 there exists an open trapping region U for ϕ^t such that $A = \bigcap_{t \geq 0} \phi^t(U)$. Because U is a trapping region for ϕ^t there exists a positive real number T such that $\phi^T(\mathrm{cl}(U)) \subseteq \mathrm{int}(U)$.

Let $\epsilon > 0$. Without loss of generality, we may assume that

$$\epsilon < D(\phi^T(\mathrm{cl}(U)), X \setminus U)$$

where D is the Hausdorff metric on the collection of nonempty closed bounded subsets of X. Since x and y are chain equivalent with respect to ϕ^t there exists an (ϵ, T)-chain

$$(x = x_0, x_1, ..., x_n = y; t_0, t_1, ..., t_{n-1})$$

from x to y. If $x_i \in U$ for some $i = 0, \ldots, n-1$, then $\phi^{t_i}(x_i) \in \phi^{t_i}(U) \subseteq \phi^T(\mathrm{cl}(U))$. Since

$$d(x_{i+1}, \phi^{t_i}(x_i)) < \epsilon < D(\phi^T(\mathrm{cl}(U)), X \setminus U),$$

we conclude $x_{i+1} \in U$. Therefore, if $x \in A \subseteq U$, then $y \in U$. Since x and y are chain equivalent with respect to ϕ^t, they are chain recurrent with respect to ϕ^t. Consequently, $y \in A$ by Lemma 3.4.26. If $x \in A^*$, then $y \in A^*$ since otherwise the preceding argument applied to $y \in A$ implies that $x \in A$, contradicting the disjointness of A and A^*.

To prove the converse by means of contraposition, assume that $x, y \in X$ are not chain equivalent with respect to ϕ^t. If either x or y is not chain recurrent with respect to ϕ^t, then there is an attracting-repelling pair (A, A^*) for ϕ^t such that either $x \notin A \cup A^*$ or $y \notin A \cup A^*$ by Theorem 3.4.27. In particular, neither $x, y \in A$ nor $x, y \in A^*$.

So, we may assume that x and y are both chain recurrent with respect to ϕ^t. Since x, y are not chain equivalent there exist $\epsilon > 0$ and $T > 0$ so that either there is no (ϵ, T)-chain from x to y or there is no (ϵ, T)-chain from y to x. Without loss of generality, we may assume that there is no (ϵ, T)-chain from x to y. Let

$$U = \{z \in X \mid \forall s \in [0, T] \text{ there is an } (\epsilon, T)\text{-chain from } x \text{ to } \phi^{-s}(z)\}.$$

According to Example 3.4.7, the set U is a trapping region for ϕ^t such that $x \in U$ and $y \in X \setminus U$. By Theorem 3.4.27, $x, y \in A_U \cup A_U^*$ where (A_U, A_U^*) is the attracting-repelling pair for ϕ^t corresponding to U. Since x is chain recurrent with respect to ϕ^t and $x \in U$, Lemma 3.4.26 implies that $x \in A_U$. But, $A_U \subseteq U$ and $y \in X \setminus U$. Therefore, $y \in A_U^*$. □

3.5. Exercises

(1) Let ϕ^t be a flow on a Hausdorff topological space. Prove that the equivalence relation $=$ on $\mathrm{Fix}(\phi^t)$ induces a partition of $\mathrm{Fix}(\phi^t)$ into closed invariant sets. This is the Decomposition Property of the fixed set. See Proposition 2.1.10.

(2) Let ϕ^t be a flow. Define a relation $\sim$ on $\mathrm{Per}(\phi^t)$ by

$$x \sim y \text{ if and only if } y \in \mathcal{O}(x).$$

(a) Prove that $\sim$ is an equivalence relation.
(b) Prove that $\sim$ induces a partition of $\mathrm{Per}(\phi^t)$ into closed invariant sets.

This is the Decomposition Property of the periodic set. See Proposition 2.2.18.

(3) Find all subsets of S^1 which are minimal with respect to the arrested rotation.

(4) Let ϕ^t be a flow on a first-countable Hausdorff topological space X. Prove that a nonempty compact subset A of X is minimal with respect to ϕ^t if and only if $\alpha(x) = A$ for all $x \in A$.

(5) Provide an example of a flow ϕ^t such that $M(\phi^t)$ is not closed.

(6) Prove that if ϕ^t is a flow, then $M(\phi^t)$ is invariant with respect to ϕ^t.

(7) Prove Theorem 3.2.10 (ii).

(8) Prove Corollary 3.2.15 without using Proposition 3.2.14.

(9) (a) Prove that a compact minimal set with respect to a flow ϕ^t on a metric space is chain transitive with respect to ϕ^t.
   (b) Provide an example of a minimal set A with respect to a flow ϕ^t on a metric space such that A is not chain transitive with respect to ϕ^t.

(10) Let ϕ^t be a flow on a compact second-countable Hausdorff topological space X. Determine whether the collection

$$\{\mathrm{cl}(\mathcal{O}(x)) \cap R \mid x \in X\}$$

is a partition of R into nonempty closed invariant subsets for each of the following choices of R.
   (a) $R = \mathrm{Rec}(\phi^t)$,
   (b) $R = \mathrm{cl}(\mathrm{Rec}(\phi^t))$,
   (c) $R = C(\phi^t)$,
   (d) $R = \Omega(\phi^t)$.

This exercise points out some difficulties in using orbit closures to decompose sets of recurrent points.

(11) Let ϕ^t be a flow on a compact second-countable Hausdorff topological space X. Determine whether the collection

$$\{(\alpha(x) \cup \omega(x)) \cap R \mid x \in X\}$$

is a partition of R into nonempty closed invariant subsets for each of the following choices of R.
   (a) $R = \mathrm{Rec}(\phi^t)$,
   (b) $R = \mathrm{cl}(\mathrm{Rec}(\phi^t))$,
   (c) $R = C(\phi^t)$,
   (d) $R = \Omega(\phi^t)$.

This exercise points out some difficulties in using limit sets to decompose sets of recurrent points.

(12) Provide an example of a flow ϕ^t on a metric space X and a set $A \subseteq X$ such that A is nonempty and for each $x, y \in A$ and for each $\epsilon > 0$ and $T > 0$ there exists an (ϵ, T)-chain from x to y, but A is neither closed nor invariant with respect to ϕ^t.

(13) Prove or disprove: If ϕ^t is a flow on a metric space and A is chain transitive with respect to ϕ^t, then $\phi^t(A)$ is chain transitive with respect to ϕ^t for all real numbers t.

(14) Let X and Y be compact metric spaces. Prove that if $h : X \to Y$ is a topological conjugacy from the flow $\phi^t : X \to X$ to the flow $\psi^t : Y \to Y$ and A is a chain transitive set with respect to ϕ^t, then $h(A)$ is a chain transitive set with respect to ψ^t.

(15) Provide an example of a flow ϕ^t on a nonempty topological space X and a point $x \in X$ such that neither $\alpha(x)$ nor $\omega(x)$ is topologically transitive.

(16) Provide an example of a flow ϕ^t with a positively invariant set A such that the interior of A is nonempty and A is not a trapping region for ϕ^t.

(17) Prove or disprove: If U is a trapping region for a flow ϕ^t, then there exists a positive real number T such that

$$\bigcup_{t \geq T} \phi^t(U) \subseteq \operatorname{int}(U).$$

(18) Define $f : [-2/(3\pi), 2/\pi] \to \mathbb{R}$ by

$$f(x) = \begin{cases} x^2 \sin \dfrac{1}{x} & \text{if } x \neq 0, \\ 0 & \text{if } x = 0. \end{cases}$$

(a) Prove that there exists a differentiable periodic function $g : \mathbb{R} \to \mathbb{R}$ with period 2π such that $g|_{[-2/(3\pi),2/\pi]} = f$ and the zeros of g are 0 and $1/(n\pi)$ for each integer n.

(b) Prove that the flow of the differential equation

$$\dot{x} = g(x)$$

on S^1 has countably many attracting-repelling pairs.

(19) Let A be an $n \times n$ matrix with real entries such that the real part of each eigenvalue of A is positive. Let ϕ^t be the flow of the linear differential equation

$$\dot{x} = Ax$$

on $\mathbb{R}^n$. Prove that $\{0\}$ is a repeller for ϕ^t.

(20) Let ϕ^t be the arrested logistic rotation-dilation on the closed unit disk $\mathbb{D}$. Prove that $\partial\mathbb{D}$ is an attractor for ϕ^t and $\{0\}$ is a repeller for ϕ^t.

(21) Provide an example of a flow with a chain transitive set which is neither an attracting set nor a repelling set, hence neither an attractor nor a repeller.

(22) (a) Let A be a nonempty attracting set for a flow on a topological space X. Prove that if there exists $x \in X$ such that $\omega(x) = A$, then A is an attractor.
(b) Let A^* be a nonempty repelling set for a flow on a topological space X. Prove that if there exists $x \in X$ such that $\alpha(x) = A^*$, then A^* is a repeller.

(23) Consider the flow ϕ^t corresponding to the system of differential equations

$$\begin{aligned} \dot{r} &= r(1-r), \\ \dot{\theta} &= 0 \end{aligned}$$

in polar coordinates on $\mathbb{C}$.
(a) Prove that the unit circle $\{z \in \mathbb{C} \mid |z| = 1\}$ is an attractor for ϕ^t.
(b) Prove that there does not exist $z \in \mathbb{C}$ such that $\omega(z)$ is the unit circle.

(24) Consider the flow ϕ^t of the system of differential equations

$$\begin{aligned} \dot{r} &= r(1-r^2), \\ \dot{\theta} &= \theta^2(2\pi - \theta)^2 \end{aligned}$$

in polar coordinates on $\mathbb{C}$.
(a) Let $\epsilon \in (0,1)$. Let $U = \{z \in \mathbb{C} \mid ||z| - 1| < \epsilon\}$. Prove that $\omega(z) = \{1\}$ for all $z \in U$.
(b) Prove that U is a trapping region for ϕ^t.
(c) Prove that $A_U = \{z \in \mathbb{C} \mid |z| = 1\}$.

(25) Let ϕ^t be a flow on a compact metric space X. Let A be a compact invariant set with respect to ϕ^t. Prove that A is an attracting set if and only if for every $\delta > 0$ there exists a positively invariant closed bounded subset V of X such that $A \subseteq \operatorname{int}(V)$, $D(V, A) < \delta$, and $\omega(x) \subseteq A$ for all $x \in V$, where D is the Hausdorff metric on the collection of all closed bounded subsets of X.

(26) A subset A of $\mathbb{R}$ is **syndetic** if there exists a compact subset K of $\mathbb{R}$ such that $\mathbb{R} = A + K$. Let ϕ^t be a flow on a topological space X. A point $x \in X$ is **almost periodic** with respect to ϕ^t if for each open neighborhood U of x in X the set $\{t \mid \phi^t(x) \in U\}$ is syndetic.
(a) Let ϕ^t be a flow on a locally compact Hausdorff topological space X. Prove that $x \in X$ is almost periodic with respect to ϕ^t if and only if $\operatorname{cl}(\mathcal{O}(x))$ is a compact subset of X which is minimal with respect to ϕ^t.

(b) Let ϕ^t be a flow on a topological space X. Prove that every nonempty compact Hausdorff invariant subset of X contains an almost periodic point with respect to ϕ^t.

(c) Let ϕ^t be a flow on a locally compact Hausdorff topological space X. Prove that every point of X is almost periodic with respect to ϕ^t if and only if $\{\mathrm{cl}(\mathcal{O}(x)) \mid x \in x\}$ is a partition of X consisting of compact sets.

Chapter 4

Test Functions

This chapter examines real-valued functions defined on the phase space of a flow that reflect the flow's behavior. We refer to such real-valued functions as test functions. A test function often provides valuable information about a flow, especially a flow arising from a system of differential equations that does not readily admit an analytical solution.

Two types of test functions are important, those which are constant along orbits and those which decrease along orbits. Hamiltonian functions and their generalizations, such as invariant functions, are test functions of the first type. Potential functions and their generalizations, such as weak and strong Lyapunov functions, are test functions of the second type. Morse functions are particular potential functions which motivate much of the theory. Complete Lyapunov functions combine the salient features of the two types of test functions; they are constant on each chain component and strictly decrease along orbits in the complement of the chain recurrent set.

According to the Fundamental Theorem of Dynamical Systems due to Conley [**11**] every flow on a compact metric space has a complete Lyapunov function. This chapter concludes with a statement, proof and interpretation of the Fundamental Theorem of Dynamical Systems.

4.1. Potential Functions on Euclidean Spaces and Surfaces

In this section we study gradient flows on Euclidean spaces and surfaces in $\mathbb{R}^3$. Section 4.5 explores gradient flows in the more general context of manifolds. We begin by reviewing the gradient vector and its most important properties.

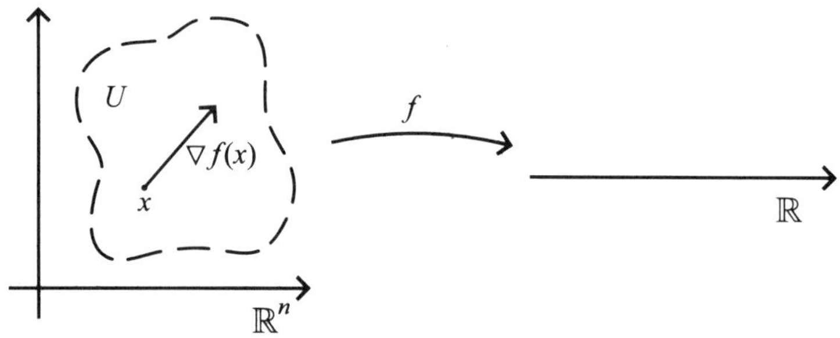

Figure 4.1. The gradient vector

Definition 4.1.1. Let U be an open subset of $\mathbb{R}^n$. If $f : U \to \mathbb{R}$ is differentiable, then the **gradient vector** of f at $x \in U$ is the vector

$$\nabla f(x) = \begin{pmatrix} \dfrac{\partial f}{\partial x_1}(x) \\ \vdots \\ \dfrac{\partial f}{\partial x_n}(x) \end{pmatrix}.$$

If $f : U \to \mathbb{R}$ is differentiable, then f increases most rapidly at x in the direction of $\nabla f(x)$, while f decreases most rapidly at x in the direction of $-\nabla f(x)$. See Figure 4.1.

Consider a particle moving in an open subset U of $\mathbb{R}^n$ with position function $x : \mathbb{R} \to U$ with respect to time. Let $V : U \to \mathbb{R}$ be the potential energy of the particle as a function of the particle's position. If the particle always moves to decrease its potential energy most rapidly, then the differential equation

$$\dot{x} = -\nabla V(x)$$

models the particle's motion.

Definition 4.1.2. Let U be an open subset of $\mathbb{R}^n$. A vector field F on U is a **gradient vector field** if there exists a differentiable function $V : U \to \mathbb{R}$ such that

$$F = -\nabla V.$$

Definition 4.1.3. Let U be an open subset of $\mathbb{R}^n$. A differentiable flow ϕ^t on U is a **gradient flow** if there exists a gradient vector field F on U such that

$$\left.\frac{d}{dt}\right|_{t=0} \phi^t(x) = F(x)$$

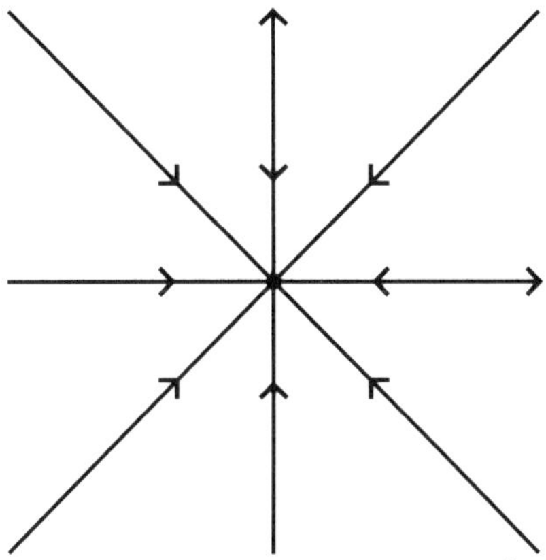

Figure 4.2. A phase portrait of the flow of Example 4.1.4

for all $x \in U$, in which case there exists a differentiable function $V : U \to \mathbb{R}$ such that

$$\left.\frac{d}{dt}\right|_{t=0} \phi^t(x) = -\nabla V(x)$$

for all $x \in U$. The function V is a **potential function** for the vector field F or for the flow ϕ^t.

Example 4.1.4. Consider the flow ϕ^t on $\mathbb{R}^2$ defined by

$$\phi^t(x, y) = e^{-t}(x, y).$$

Define $V : \mathbb{R}^2 \to \mathbb{R}$ by

$$V(x, y) = \frac{1}{2}(x^2 + y^2).$$

The flow ϕ^t is a gradient flow, and V is a potential function for ϕ^t since

$$\left.\frac{d}{dt}\right|_{t=0} \phi^t(x, y) = -\begin{pmatrix} x \\ y \end{pmatrix} = -\nabla V(x, y).$$

See Figure 4.2.

Not every flow is a gradient flow.

Proposition 4.1.5. *Let U be an open subset of $\mathbb{R}^2$. Let $F : U \to \mathbb{R}^2$ be a C^1 vector field on U with component functions F_1 and F_2. If F is a gradient vector field, then*

$$\frac{\partial F_1}{\partial y} = \frac{\partial F_2}{\partial x}.$$

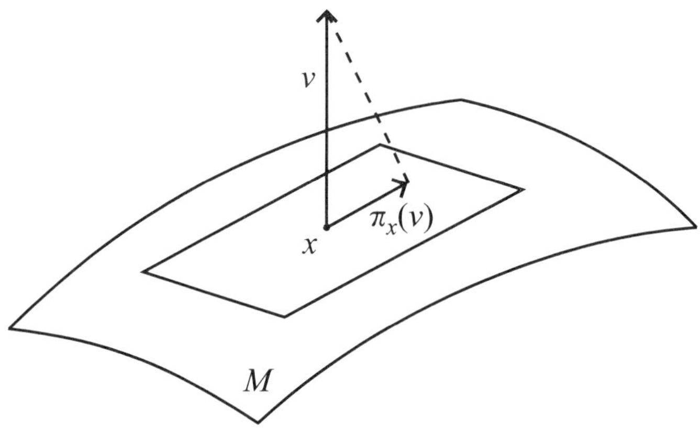

Figure 4.3. Orthogonal projection into a tangent plane to a surface

Proof. If F is a C^1 gradient vector field, then there exists a C^2 potential function $V : U \to \mathbb{R}$ for F such that

$$\frac{\partial V}{\partial x} = F_1 \quad \text{and} \quad \frac{\partial V}{\partial y} = F_2.$$

Thus,

$$\frac{\partial^2 V}{\partial y\, \partial x} = \frac{\partial F_1}{\partial y} \quad \text{and} \quad \frac{\partial^2 V}{\partial x\, \partial y} = \frac{\partial F_2}{\partial x}.$$

By the equality of mixed partial derivatives for C^2 functions,

$$\frac{\partial F_1}{\partial y} = \frac{\partial F_2}{\partial x}.$$

□

Example 4.1.6. Define a vector field $F : \mathbb{R}^2 \to \mathbb{R}^2$ by

$$F(x, y) = \begin{pmatrix} y \\ -x \end{pmatrix}.$$

By Proposition 4.1.5 the vector field F is not a gradient vector field because

$$\frac{\partial}{\partial y} y = 1 \neq -1 = \frac{\partial}{\partial x}(-x).$$

Proposition 4.1.5 provides a necessary condition for a vector field to be a gradient vector field. The search for sufficient conditions for a vector field to be a gradient vector field motivates much of algebraic topology.

Now, let M be a surface in $\mathbb{R}^3$. Let π_x denote the orthogonal projection of vectors in $\mathbb{R}^3$ at x to the tangent plane to M at x. See Figure 4.3.

Definition 4.1.7. Let M be a surface in $\mathbb{R}^3$. A vector field F on M is a **gradient vector field** if there exists an open neighborhood U of M in $\mathbb{R}^3$

and a differentiable function $V : U \to \mathbb{R}$ such that

$$F(x) = -\pi_x(\nabla V(x)).$$

Definition 4.1.8. Let M be a surface in $\mathbb{R}^3$. A differentiable flow ϕ^t on M is a **gradient flow** if there exists a gradient vector field F on an open neighborhood of M in $\mathbb{R}^3$ such that

$$\left.\frac{d}{dt}\right|_{t=0} \phi^t(x) = F(x)$$

for all $x \in M$, in which case there exists an open neighborhood U of M in $\mathbb{R}^3$ and a differentiable function $V : U \to \mathbb{R}$ such that

$$\left.\frac{d}{dt}\right|_{t=0} \phi^t(x) = -\pi_x(\nabla V(x))$$

for all $x \in M$. The function V is a **potential function** for the vector field F or for the flow ϕ^t.

Example 4.1.9. Let $S^2 = \{(x, y, z) \in \mathbb{R}^3 \mid x^2 + y^2 + z^2 = 1\}$. Define $V : \mathbb{R}^3 \to \mathbb{R}$ by

$$V(x, y, z) = z.$$

The function V returns the height of a point above or below the xy-plane in space, and

$$\nabla V(x, y, z) = \begin{pmatrix} 0 \\ 0 \\ 1 \end{pmatrix}.$$

Projecting ∇V to the tangent planes to S^2 we obtain

$$\pi_{(x,y,z)}(\nabla V(x, y, z)) = \begin{pmatrix} -xz \\ -yz \\ x^2 + y^2 \end{pmatrix}.$$

See Exercise 3. Thus, the system of differential equations

$$\begin{aligned} \dot{x} &= xz, \\ \dot{y} &= yz, \\ \dot{z} &= -(x^2 + y^2) \end{aligned}$$

defines a gradient flow on S^2 called the **height flow on the sphere** with potential function V. See Figure 4.4.

Example 4.1.10. A parametrization of the torus $\mathbb{T}^2$ in $\mathbb{R}^3$ is

$$\begin{aligned} x(\alpha, \theta) &= (2 + \cos\alpha)\cos\theta, \\ y(\alpha, \theta) &= -\sin\alpha, \\ z(\alpha, \theta) &= (2 + \cos\alpha)\sin\theta \end{aligned}$$

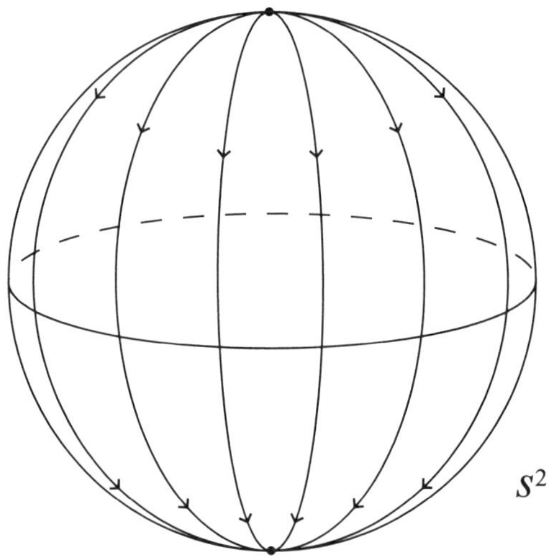

Figure 4.4. A phase portrait of the height flow on the sphere

where $\alpha, \theta \in [0, 2\pi]$. Define $V : \mathbb{R}^3 \to \mathbb{R}$ by

$$V(x, y, z) = z.$$

As in Example 4.1.9 the function V returns the height of a point above or below the xy-plane in space, and

$$\nabla V(x, y, z) = \begin{pmatrix} 0 \\ 0 \\ 1 \end{pmatrix}.$$

Projecting ∇V to the tangent planes to $\mathbb{T}^2$ we obtain

$$\pi_{(\alpha,\theta)}(\nabla V(\alpha, \theta)) = \begin{pmatrix} -\cos^2\alpha \sin\theta\cos\theta \\ \sin\alpha\cos\alpha\sin\theta \\ \sin^2\alpha\sin^2\theta + \cos^2\theta \end{pmatrix}.$$

See Exercise 4. Thus, the system of differential equations

$$\begin{aligned} \dot{\alpha} &= \sin\alpha\sin\theta, \\ \dot{\theta} &= \frac{-\cos\theta}{2 + \cos\alpha} \end{aligned}$$

defines a gradient flow on $\mathbb{T}^2$ called the **height flow on the torus** with potential function V. See Figure 4.5.

Perhaps the most important fact about gradient flows and potential functions is that a potential function decreases along the orbits of its gradient flow.

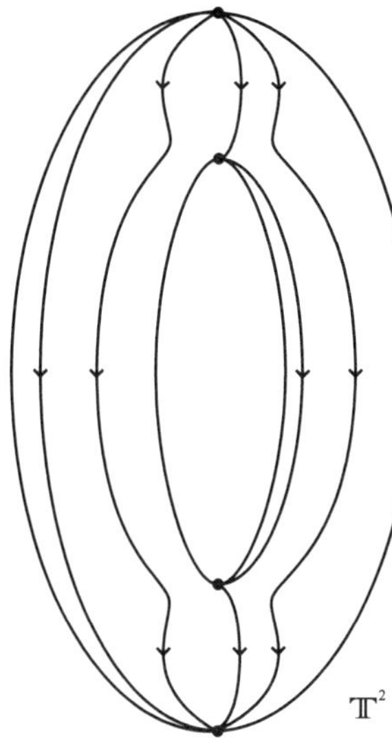

Figure 4.5. A phase portrait of the height flow on the torus

Proposition 4.1.11. *If ϕ^t is a gradient flow on a compact surface $M \subset \mathbb{R}^3$ (respectively, an open subset U of $\mathbb{R}^n$) with a potential function $V : M \to \mathbb{R}$ (respectively, $V : U \to \mathbb{R}$), then $V(\phi^t(x)) \leq V(x)$ for all $x \in M$ (respectively, $x \in U$) and all $t > 0$ with equality if and only if x is a fixed point of ϕ^t.*

Proof. We will prove the proposition for a gradient flow ϕ^t on a compact surface $M \subset \mathbb{R}^3$. Let $x \in M$. By the Chain Rule,

$$\begin{aligned} \left.\frac{d}{dt}\right|_{t=0} V(\phi^t(x)) &= \nabla V(\phi^0(x)) \cdot \left.\frac{d}{dt}\right|_{t=0} \phi^t(x) \\ &= \nabla V(x) \cdot -\pi_x(\nabla V(x)) \\ &= -\|\nabla V(x)\| \, \|\pi_x(\nabla V(x))\| \cos\theta \\ &\leq 0 \end{aligned}$$

where θ is the angle between the vectors $\nabla V(x)$ and $\pi_x(\nabla V(x))$. Since the angle between a vector and its projection is acute, the cosine of θ is nonnegative. Therefore, $V(\phi^t(x)) < V(x)$ for all $x \in M$ and $t > 0$. Finally, $V(\phi^t(x)) = V(x)$ for all $t > 0$ if and only if

$$\frac{d}{dt} V(\phi^t(x)) = 0$$

if and only if $\pi_x(\nabla V(x)) = 0$ if and only if x is a fixed point of ϕ^t. The proof for a gradient flow on an open subset of $\mathbb{R}^n$ is similar. $\square$

The monotonicity of a potential function along the orbits of its gradient flow forces every orbit to converge to a set of fixed points.

Proposition 4.1.12. *If ϕ^t is a gradient flow on a compact surface $M \subset \mathbb{R}^3$, then $\omega(x) \subseteq Fix(\phi^t)$ and $\alpha(x) \subseteq Fix(\phi^t)$ for all $x \in M$.*

Proof. Since ϕ^t is a gradient flow on M, there exists a neighborhood U of M and a potential function $V : U \to \mathbb{R}$ for ϕ^t. Let $x \in M$. By means of contradiction assume that $y \in \omega(x)$ is not a fixed point of ϕ^t. Let $\tau > 0$. Define

$$\delta = V(y) - V(\phi^\tau(y)).$$

By Proposition 4.1.11, $V(\phi^t(y))$ is a strictly decreasing function of t. Thus, $\delta > 0$. Since $y \in \omega(x)$ there exists a sequence $t_n \to \infty$ such that

$$\lim_{n\to\infty} \phi^{t_n}(x) = y.$$

By the group property of flows and the continuity of the functions ϕ^t and V we obtain

$$\begin{aligned} \lim_{n\to\infty} V(\phi^{\tau+t_n}(x)) &= \lim_{n\to\infty} V(\phi^\tau(\phi^{t_n}(x))) \\ &= V(\phi^\tau(y)) \\ &= V(y) - \delta. \end{aligned}$$

So, there exists $N > 0$ such that

$$V(\phi^{\tau+t_n}(x)) < V(y) - \delta + \frac{\delta}{2} = V(y) - \frac{\delta}{2}$$

for all $n \geq N$, By Proposition 4.1.11 the function V decreases along orbits. Consequently,

$$V(\phi^t(x)) < V(y) - \frac{\delta}{2}$$

for all $t > t_N$.

On the other hand, since $\lim_{n\to\infty} \phi^{t_n}(x) = y$, the continuity of V guarantees that

$$\lim_{n\to\infty} V(\phi^{t_n}(x)) = V(y).$$

This is a contradiction. So, y is a fixed point of ϕ^t. Therefore, $\omega(x) \subseteq \mathrm{Fix}(\phi^t)$ for all $x \in M$. The proof that $\alpha(x) \subseteq \mathrm{Fix}(\phi^t)$ for all $x \in M$ is similar. □

Gradient flows provide an excellent context in which to apply Theorem 3.4.27 to find the chain recurrent set of a flow. In preparation for this application we introduce stable and unstable manifolds.

Definition 4.1.13. Let ϕ^t be a flow on a metric space (X, d). For each $x \in X$ the **stable manifold** of x is the set

$$W^s(x) = \{y \in X \mid d(\phi^t(x), \phi^t(y)) \to 0 \text{ as } t \to \infty\},$$

and the **unstable manifold** of x is the set

$$W^u(x) = \{y \in X \mid d(\phi^t(x), \phi^t(y)) \to 0 \text{ as } t \to -\infty\}.$$

Remark 4.1.14. The use of the term *manifold* in this setting is unfortunate since the stable manifold or unstable manifold of a point might not be a *bona fide* submanifold of the phase space. Despite this infelicity we retain the terms stable manifold and unstable manifold for historical reasons and because the more natural terms stable set and unstable set have other meanings in the literature.

Theorem 3.4.27 allows us to find the chain recurrent set of the height flow on the torus.

Example 4.1.15. Recall from Example 4.1.10 that

$$\begin{aligned} x(\alpha, \theta) &= (2 + \cos\alpha)\cos\theta, \\ y(\alpha, \theta) &= -\sin\alpha, \\ z(\alpha, \theta) &= (2 + \cos\alpha)\sin\theta \end{aligned}$$

where $\alpha, \theta \in [0, 2\pi]$ is a parametrization of the torus $\mathbb{T}^2$ in $\mathbb{R}^3$. Example 4.1.10 showed that the system of differential equations

$$\begin{aligned} \dot\alpha &= \sin\alpha\sin\theta, \\ \dot\theta &= \frac{-\cos\theta}{2 + \cos\alpha} \end{aligned}$$

defines a gradient flow ϕ^t on $\mathbb{T}^2$. The fixed points of ϕ^t are

$$p_0 = \left(0, \frac{3\pi}{2}\right), \; p_1 = \left(\pi, \frac{3\pi}{2}\right), \; p_2 = \left(\pi, \frac{\pi}{2}\right) \text{ and } p_3 = \left(0, \frac{\pi}{2}\right).$$

The stable and unstable manifolds of p_1 and p_2 are

$$W^s(p_1) = \left\{(\pi, \theta) \mid \theta \neq \frac{\pi}{2}\right\},$$

$$W^u(p_1) = \left\{\left(\alpha, \frac{3\pi}{2}\right) \mid \alpha \neq 0\right\},$$

$$W^s(p_2) = \left\{\left(\alpha, \frac{\pi}{2}\right) \mid \alpha \neq 0\right\}$$

and

$$W^u(p_2) = \left\{(\pi, \theta) \mid \theta \neq \frac{3\pi}{2}\right\}.$$

If $U_0 = \emptyset$, then U_0 is a trapping region such that

$$A_{U_0} = \emptyset \text{ and } A^*_{U_0} = \mathbb{T}^2.$$

If U_1 is a trapping region containing p_0, but not containing any other fixed points, then

$$A_{U_1} = \{p_0\} \text{ and } A^*_{U_1} = \{p_3\} \cup W^s(p_1) \cup W^s(p_2).$$

If U_2 is a trapping region containing p_0 and p_1, but not p_2 or p_3, then

$$A_{U_2} = \{p_0\} \cup W^u(p_1) \text{ and } A^*_{U_2} = \{p_3\} \cup W^s(p_2).$$

If U_3 is a trapping region containing p_0 and p_2, but not p_1 or p_3, then

$$A_{U_3} = \{p_0\} \cup W^u(p_2) \text{ and } A^*_{U_3} = \{p_3\} \cup W^s(p_1).$$

If U_4 is a trapping region containing p_0, p_1 and p_2, but not p_3, then

$$A_{U_4} = \{p_0\} \cup W^u(p_1) \cup W^u(p_2) \text{ and } A^*_{U_4} = \{p_3\}.$$

If $U_5 = \mathbb{T}^2$, then U_5 is a trapping region such that

$$A_{U_5} = \mathbb{T}^2 \text{ and } A^*_{U_5} = \emptyset.$$

See Figure 4.6.

By Theorem 3.4.27 and Example 2.7.5,

$$\begin{aligned} \mathcal{R}(\phi^t) &= \bigcap \{A_U \cup A^*_U \mid U \text{ is a trapping region for } \phi^t\} \\ &\subseteq \bigcap_{i=0}^{5} (A_{U_i} \cup A^*_{U_i}) \\ &= \{p_0, p_1, p_2, p_3\} \\ &= \operatorname{Fix}(\phi^t) \\ &\subseteq \mathcal{R}(\phi^t). \end{aligned}$$

Therefore,

$$\mathcal{R}(\phi^t) = \operatorname{Fix}(\phi^t).$$

4.2. Hamiltonian Functions

Consider a particle with mass m moving in an open subset U of $\mathbb{R}^n$ with position $q : \mathbb{R} \to U$ and potential energy $V : U \to \mathbb{R}$ according to Newton's Second Law of Motion

$$m\ddot{q} = -\nabla V(q).$$

The particle's momentum $p : \mathbb{R} \to \mathbb{R}^n$ is

$$p = m\dot{q}.$$

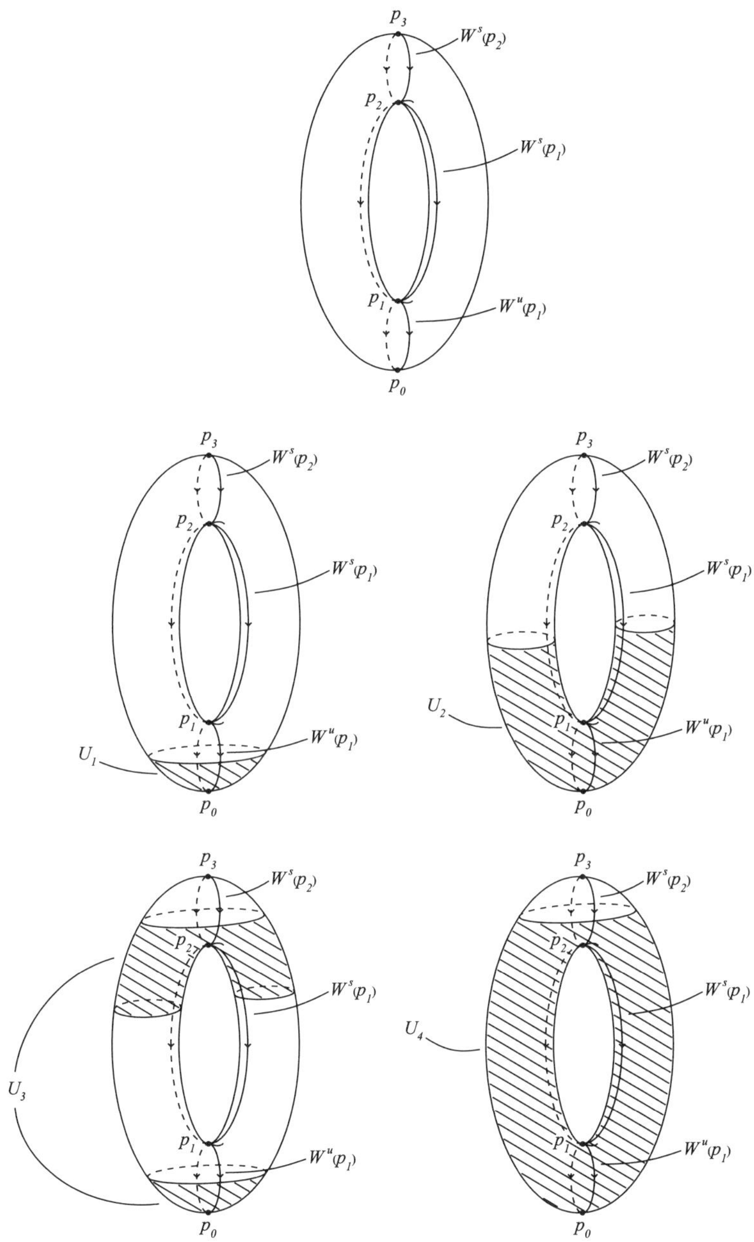

Figure 4.6. Trapping regions for the height flow on the torus

Let $q_1, \ldots, q_n$ and $p_1, \ldots, p_n$ be the component functions of q and p, respectively. Thus,

$$\begin{aligned} \dot{q}_j &= \frac{p_j}{m}, \\ \dot{p}_j &= -\frac{\partial V}{\partial q_j} \end{aligned}$$

for $j = 1, \ldots, n$.

The particle's kinetic energy is

$$\frac{1}{2} m \|\dot{q}\|^2 = \frac{1}{2} m (\dot{q} \cdot \dot{q}) = \frac{p \cdot p}{2m}.$$

Therefore, the total energy of the particle in the configuration (q, p) is the sum of its potential and kinetic energies:

$$V(q) + \frac{p \cdot p}{2m}.$$

The space of all possible ordered pairs of positions and momenta of the particle is $U \times \mathbb{R}^n$. Define $H : U \times \mathbb{R}^n \to \mathbb{R}$ by

$$H(q, p) = V(q) + \frac{p \cdot p}{2m}.$$

Expressing Newton's Second Law of Motion in terms of H we obtain **Hamilton's differential equations**

$$\begin{aligned} \dot{q}_j &= \frac{\partial H}{\partial p_j}, \\ \dot{p}_j &= -\frac{\partial H}{\partial q_j} \end{aligned}$$

for $j = 1, \ldots, n$, named for the Irish mathematician William Rowan Hamilton (1805–1865). We can express Hamilton's differential equations via the gradient of H by introducing the $2n \times 2n$ matrix

$$J_n = \begin{pmatrix} 0_n & I_n \\ -I_n & 0_n \end{pmatrix}$$

where 0_n is the $n \times n$ zero matrix and I_n is the $n \times n$ identity matrix. Regarding $\nabla H(x)$ as a $2n$-vector, Hamilton's differential equations have the form

$$\dot{x} = J_n \nabla H(x)$$

where $x = (q_1, \ldots, q_n, p_1, \ldots, p_n)$.

Definition 4.2.1. Let U be an open subset of $\mathbb{R}^n$. A vector field F on $U \times \mathbb{R}^n$ is a **Hamiltonian vector field** if there exists a differentiable function $H : U \times \mathbb{R}^n \to \mathbb{R}$ such that

$$F(x) = J_n \nabla H(x)$$

for all $x \in U \times \mathbb{R}^n$.

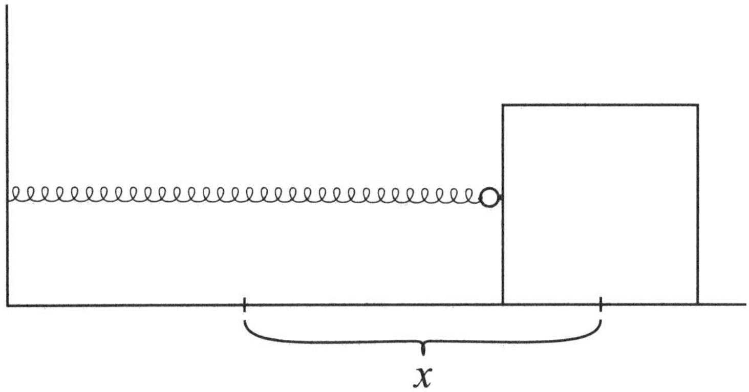

Figure 4.7. A harmonic oscillator

Definition 4.2.2. Let U be an open subset of $\mathbb{R}^n$. A smooth flow ϕ^t on $U \times \mathbb{R}^n$ is a **Hamiltonian flow** if there exists a Hamiltonian vector field F such that

$$\left.\frac{d}{dt}\right|_{t=0} \phi^t(x) = F(x)$$

for all $x \in U \times \mathbb{R}^n$, in which case there exists a differentiable function $H : U \times \mathbb{R}^n \to \mathbb{R}$ such that

$$\left.\frac{d}{dt}\right|_{t=0} \phi^t(x) = J_n \nabla H(x)$$

for all $x \in U \times \mathbb{R}^n$. The function H is a **Hamiltonian function** for the vector field F or for the flow ϕ^t.

Example 4.2.3. Consider the flow corresponding to the system of differential equations

$$\begin{aligned} \dot{x} &= y, \\ \dot{y} &= -x \end{aligned}$$

describing the motion of an (undamped) **harmonic oscillator** with displacement x from equilibrium and velocity y. See Figure 4.7. The harmonic oscillator determines a vector field

$$F(x, y) = \begin{pmatrix} y \\ -x \end{pmatrix}$$

on $\mathbb{R}^2$.

The total energy function for a harmonic oscillator is the function $H : \mathbb{R} \times \mathbb{R} \to \mathbb{R}$ defined by

$$H(x, y) = \frac{1}{2}(x^2 + y^2)$$

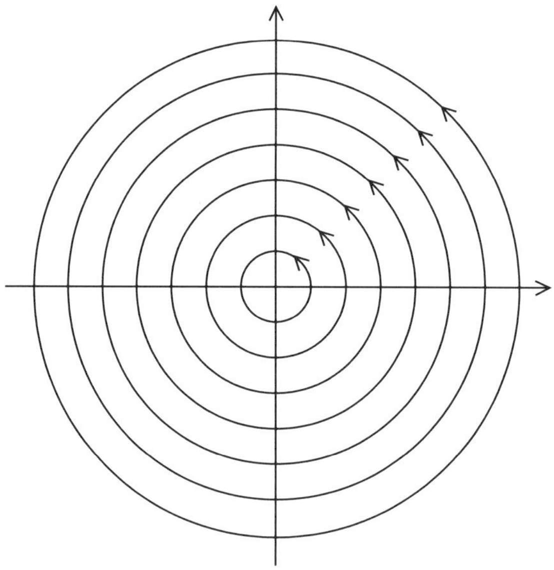

Figure 4.8. A phase portrait of a harmonic oscillator

with gradient

$$\nabla H(x, y) = \begin{pmatrix} x \\ y \end{pmatrix}.$$

So,

$$J_2 \nabla H(x, y) = \begin{pmatrix} 0 & 1 \\ -1 & 0 \end{pmatrix} \begin{pmatrix} x \\ y \end{pmatrix} = \begin{pmatrix} y \\ -x \end{pmatrix} = F(x, y).$$

Therefore, the vector field F for the harmonic oscillator is Hamiltonian, and H is a corresponding Hamiltonian function.

By solving the system we obtain the flow ϕ^t on $\mathbb{R} \times \mathbb{R}$ defined by

$$\phi^t(x, y) = (x \cos t + y \sin t, -x \sin t + y \cos t).$$

See Figure 4.8. Differentiating with respect to t we obtain

$$\left.\frac{d}{dt}\right|_{t=0} \phi^t(x, y) = \begin{pmatrix} y \\ -x \end{pmatrix} = F(x, y).$$

Since F is a Hamiltonian vector field, ϕ^t is a Hamiltonian flow.

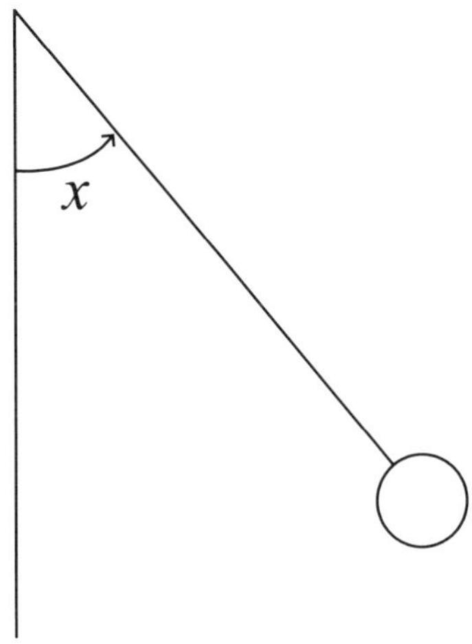

Figure 4.9. An ideal pendulum

Finally, we compute the derivative of H along a solution $(x(t), y(t))$. By the Chain Rule,

$$\begin{aligned}\left.\frac{d}{dt}\right|_{t=0} H(x(t), y(t)) &= x(0)\dot{x}(0) + y(0)\dot{y}(0) \\ &= x(0)y(0) - y(0)x(0) \\ &= 0.\end{aligned}$$

Therefore, H is constant on each orbit of ϕ^t.

Example 4.2.4. Consider the flow on $\mathbb{R} \times \mathbb{R}$ associated to the system of differential equations

$$\begin{aligned}\dot{x} &= y, \\ \dot{y} &= -\sin x\end{aligned}$$

describing the motion of an **ideal pendulum** where x is the pendulum's angular displacement from a downward equilibrium, and $y = \dot{x}$ is its angular velocity. See Figure 4.9.

The total energy function for the ideal pendulum is the function $H : \mathbb{R} \times \mathbb{R} \to \mathbb{R}$ defined by

$$H(x, y) = \frac{1}{2}y^2 + 1 - \cos x.$$

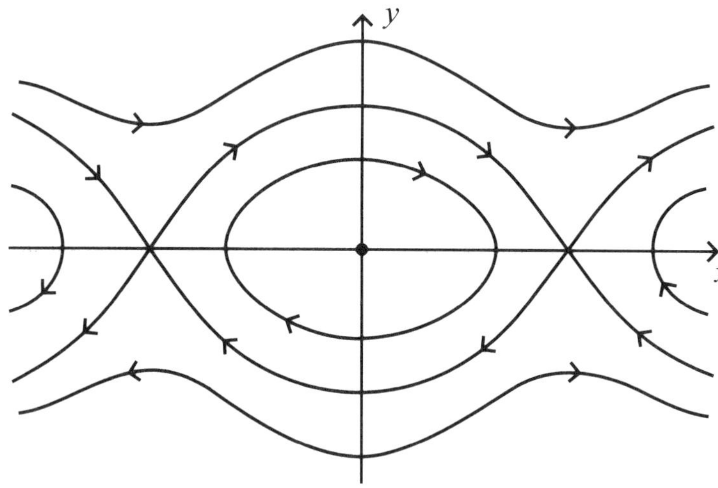

Figure 4.10. A phase portrait of an ideal pendulum

The gradient of H is

$$\nabla H(x, y) = \begin{pmatrix} -\sin x \\ y \end{pmatrix}.$$

Since

$$J_2 \nabla H(x, y) = \begin{pmatrix} 0 & 1 \\ -1 & 0 \end{pmatrix} \begin{pmatrix} -\sin x \\ y \end{pmatrix} = \begin{pmatrix} y \\ -\sin x \end{pmatrix},$$

the vector field for the ideal pendulum is Hamiltonian, the corresponding flow is Hamiltonian, and H is a Hamiltonian function.

Computing the derivative of H along a solution $(x(t), y(t))$,

$$\begin{aligned} \left.\frac{d}{dt}\right|_{t=0} H(x(t), y(t)) &= y(0)\dot{y}(0) + \sin(x(0))\dot{x}(0) \\ &= y(0)(-\sin x(0)) + \sin(x(0))y(0) \\ &= 0. \end{aligned}$$

Therefore, H is constant along orbits of the flow. See Figure 4.10.

Examples 4.2.3 and 4.2.4 illustrate an important feature of Hamiltonian functions: A Hamiltonian function is constant on each orbit of its corresponding Hamiltonian flow.

Proposition 4.2.5. *Let U be an open subset of $\mathbb{R}^n$. If ϕ^t is a Hamiltonian flow on $U \times \mathbb{R}^n$, then*

$$\left.\frac{d}{dt}\right|_{t=0} H(\phi^t(q, p)) = 0$$

for all $(q, p) \in U \times \mathbb{R}^n$.

Proof. By the Chain Rule and Hamilton's differential equations,

$$\begin{aligned}
\left.\frac{d}{dt}\right|_{t=0} H(\phi^t(q,p)) &= \sum_{j=1}^{n} \frac{\partial H}{\partial q_j}(q,p)\dot{q}_j(0) + \frac{\partial H}{\partial p_j}(q,p)\dot{p}_j(0) \\
&= \sum_{j=1}^{n} \frac{\partial H}{\partial q_j}(q,p)\frac{\partial H}{\partial p_j}(q,p) - \frac{\partial H}{\partial p_j}(q,p)\frac{\partial H}{\partial q_j}(q,p) \\
&= 0.
\end{aligned}$$

□

Corollary 4.2.6. *A Hamiltonian function is constant on each orbit of its Hamiltonian flow.*

Geometrically, Corollary 4.2.6 implies that each orbit of a Hamiltonian flow is a subset of a level set of a Hamiltonian function for the flow.

The remainder of this section investigates how Hamiltonian flows preserve volumes of sets in the phase space and how volume-preservation forces widespread recurrence. First, we establish some important results from the theory of ordinary differential equations. Second, we review some facts about subsets of Euclidean spaces and Riemann integration. Third, we prove Liouville's Theorem, establishing that Hamiltonian flows preserve volume. Finally, we prove the Poincaré Recurrence Theorem which states that every point is nonwandering with respect to a volume-preserving flow on a bounded subset of Euclidean space.

Denote the set of $n \times n$ matrices with real entries by $\mathbb{R}^{n\times n}$.

Definition 4.2.7. Let $I \subseteq \mathbb{R}$ be an open interval. Let $A : I \to \mathbb{R}^{n\times n}$ be continuous. A differentiable function $M : I \to \mathbb{R}^{n\times n}$ is a **matrix solution** of the differential equation $\dot{x} = A(t)x$ provided that $\dot{M} = A(t)M$.

Following earlier work by Niels Abel (1802–1829), Joseph Liouville (1809–1882) and Mikhail Ostrogradski (1801–1861) independently established a beautiful formula relating the determinant of a matrix solution to the trace of the coefficient matrix.

Proposition 4.2.8. The Liouville-Ostrogradski Formula. *Let $I \subseteq \mathbb{R}$ be an open interval. Let $A : I \to \mathbb{R}^{n\times n}$ be continuous. If $t_0 \in I$ and M is a matrix solution of the differential equation*

$$\dot{x} = A(t)x,$$

then

$$\det(M(t)) = \det(M(t_0)) \exp\left(\int_{t_0}^{t} tr(A(s))\, ds \right)$$

for all $t \in I$.

Proof. Let $\{e_1, \ldots, e_n\}$ be the standard basis for $\mathbb{R}^n$ considered as column vectors, and denote the ij-th entry of $A(t)$ by $a_{ij}(t)$. Applying the linearity of the determinant with respect to rows we obtain

$$\begin{aligned}
\frac{d}{dt}\det(M(t)) &= \frac{d}{dt}\det(e_1^T M(t), \ldots, e_n^T M(t)) \\
&= \sum_{i=1}^{n} \det(e_1^T M(t), \ldots, e_{i-1}^T M(t), e_i^T \dot{M}(t), \\
&\qquad\qquad e_{i+1}^T M(t), \ldots, e_n^T M(t)) \\
&= \sum_{i=1}^{n} \det(e_1^T M(t), \ldots, e_{i-1}^T M(t), e_i^T A(t)M(t), \\
&\qquad\qquad e_{i+1}^T M(t), \ldots, e_n^T M(t)) \\
&= \sum_{i=1}^{n} \det(e_1^T M(t), \ldots, e_{i-1}^T M(t), \sum_{j=1}^{n} a_{ij}(t) e_j^T M(t), \\
&\qquad\qquad e_{i+1}^T M(t), \ldots, e_n^T M(t)) \\
&= \sum_{i=1}^{n}\sum_{j=1}^{n} a_{ij}(t) \det(e_1^T M(t), \ldots, e_{i-1}^T M(t), e_j^T M(t), \\
&\qquad\qquad e_{i+1}^T M(t), \ldots, e_n^T M(t)) \\
&= \sum_{i=1}^{n} a_{ii}(t) \det(M(t)) \\
&= \operatorname{tr}(A(t)) \det(M(t)).
\end{aligned}$$

Solving the linear differential equation

$$\frac{d}{dt}\det(M(t)) = \operatorname{tr}(A(t))\det(M(t))$$

for $\det(M(t))$ with the initial condition $\det(M(t_0))$ at $t = t_0$ we obtain

$$\det(M(t)) = \det(M(t_0)) \exp\left(\int_{t_0}^{t} \operatorname{tr}(A(s))\, ds\right). \qquad \square$$

Next, we develop the First Variation Equation which shows that if ϕ^t is the flow of a smooth vector field f, then the linearization of the flow ϕ^t with respect to the space variable is the flow of the linearization of f.

Denote the set of linear transformations from $\mathbb{R}^n$ to $\mathbb{R}^n$ by $L(\mathbb{R}^n, \mathbb{R}^n)$. Let U be an open subset of $\mathbb{R}^n$, and let $\phi : \mathbb{R} \times U \to \mathbb{R}^n$ have C^2 component functions $\phi_1, \ldots, \phi_n$. We do not require ϕ to be a flow. Define $D_x\phi : U \times$

$\mathbb{R}^n \to L(\mathbb{R}^n, \mathbb{R}^n)$ by

$$D_x\phi(t,x)(h) = \begin{pmatrix} \frac{\partial \phi_1}{\partial x_1}(t,x) & \cdots & \frac{\partial \phi_1}{\partial x_n}(t,x) \\ \vdots & \ddots & \vdots \\ \frac{\partial \phi_n}{\partial x_1}(t,x) & \cdots & \frac{\partial \phi_n}{\partial x_n}(t,x) \end{pmatrix} h$$

where $x = (x_1, \ldots, x_n)$ and $h \in \mathbb{R}^n$.

Define $\dfrac{\partial D_x\phi}{\partial t} : \mathbb{R} \times U \to L(\mathbb{R}^n, \mathbb{R}^n)$ by

$$\frac{\partial D_x\phi}{\partial t}(t,x)(h) = \begin{pmatrix} \frac{\partial^2 \phi_1}{\partial t\,\partial x_1}(t,x) & \cdots & \frac{\partial^2 \phi_1}{\partial t\,\partial x_n}(t,x) \\ \vdots & \ddots & \vdots \\ \frac{\partial^2 \phi_n}{\partial t\,\partial x_1}(t,x) & \cdots & \frac{\partial^2 \phi_n}{\partial t\,\partial x_n}(t,x) \end{pmatrix} h$$

for each $h \in \mathbb{R}^n$.

Theorem 4.2.9. The First Variation Equation. *Let f be a smooth vector field on an open subset U of $\mathbb{R}^n$. If $\phi : \mathbb{R} \times U \to U$ is the flow corresponding to the differential equation*

$$\dot{x} = f(x),$$

then ϕ is smooth, and

$$\frac{\partial D_x\phi}{\partial t}(t,x) = Df_{\phi^t(x)} \circ D_x\phi(t,x)$$

for each $x \in U$. Furthermore,

$$\det(D_x\phi(0,x)) = 1,$$

so that for each $(t,x) \in \mathbb{R} \times U$ every matrix representation of $D_x\phi(t,x)$ is a fundamental matrix solution of

$$\dot{y} = Df_{\phi^t(x)}y.$$

Proof. (Sketch) A proof that ϕ is smooth is beyond the scope of this text. See Lang [**28**], pp. 371–379.

Since ϕ is the smooth flow of $\dot{x} = f(x)$, we obtain

$$\frac{\partial \phi}{\partial t}(t,x) = \frac{d}{dt}\phi^t(x) = f(\phi^t(x)).$$

Notice that $\dfrac{\partial \phi}{\partial t} : \mathbb{R} \times U \to \mathbb{R}^n$.

Differentiating with respect to x,

$$D_x\frac{\partial \phi}{\partial t}(t,x) = D_x f(\phi^t(x)) = D_x(f \circ \phi)(t,x).$$

By the Chain Rule,

$$D_x(f \circ \phi)(t, x) = Df_{\phi^t(x)} \circ D_x\phi(t, x).$$

By the equality of mixed partial derivatives of C^2 functions,

$$\frac{\partial D_x\phi}{\partial t}(t, x) = D_x \frac{\partial \phi}{\partial t}(t, x).$$

Therefore,

$$\frac{\partial D_x\phi}{\partial t}(t, x) = Df_{\phi^t(x)} \circ D_x\phi(t, x).$$

Since

$$\det(D_x\phi(0, x)) = \det(I) = 1 \neq 0,$$

for each $(t, x) \in \mathbb{R} \times U$, every matrix representation of $D_x\phi(t, x)$ is a fundamental matrix solution of the linear differential equation

$$\dot{y} = Df_{\phi^t(x)}y. \qquad \square$$

To develop the concept of volume we employ several facts about subsets of $\mathbb{R}^n$ and Riemann integration.

Definition 4.2.10. Let $I \subset \mathbb{R}$ be a bounded interval with endpoints a and b such that $a \leq b$. The **length** of I is

$$l(I) = b - a.$$

Definition 4.2.11. A set $U \subset \mathbb{R}^n$ is a **rectangle** if there exist bounded intervals $I_1, \ldots, I_n$ such that

$$U = I_1 \times \cdots \times I_n,$$

in which case the **volume** of U is

$$\text{vol}(U) = l(I_1) \cdots l(I_n).$$

Definition 4.2.12. A set $A \subseteq \mathbb{R}^n$ has **measure zero** if for each $\epsilon > 0$ there exists a cover $\{U_1, U_2, U_3, \ldots\}$ of A by rectangles such that

$$\sum_{i=1}^{\infty} \text{vol}(U_i) < \epsilon.$$

Definition 4.2.13. A set $A \subseteq \mathbb{R}^n$ is called **Jordan-measurable** if A is bounded and the boundary of A has measure zero.

The Jordan-measurability of a set A is a necessary and sufficient condition for the characteristic function $\chi_A : \mathbb{R}^n \to \{0, 1\}$ defined by

$$\chi_A(x) = \begin{cases} 1 & \text{if } x \in A, \\ 0 & \text{if } x \notin A \end{cases}$$

to be Riemann-integrable on $\mathbb{R}^n$. See Spivak [**42**].

Definition 4.2.14. If A is a Jordan-measurable subset of $\mathbb{R}^n$, then the **volume** of A is

$$\mathrm{vol}(A) = \int_{\mathbb{R}^n} \chi_A.$$

The volume of a Jordan-measurable set satisfies two important properties.

- **Monotonicity**. If A and B are Jordan-measurable sets such that $A \subseteq B$, then
$$\mathrm{vol}(A) \leq \mathrm{vol}(B).$$
- **Countable additivity**. If $\{A_1, A_2, A_3, \ldots\}$ is a countable collection of pairwise-disjoint Jordan-measurable sets, then
$$\sum_{n=1}^{\infty} \mathrm{vol}(A_n) = \mathrm{vol}\left(\bigcup_{n=1}^{\infty} A_n\right).$$

Definition 4.2.15. A flow ϕ^t on a set $D \subseteq \mathbb{R}^n$ is **volume-preserving** if for every Jordan-measurable set $A \subseteq D$ and every $t \in \mathbb{R}$ the set $\phi^t(A)$ is Jordan-measurable, and

$$\mathrm{vol}(\phi^t(A)) = \mathrm{vol}(A).$$

Theorem 4.2.16. Liouville's Theorem. *A Hamiltonian flow is volume-preserving.*

Proof. Let U be an open subset of $\mathbb{R}^n$, and let ϕ^t be a Hamiltonian flow on $U \times \mathbb{R}^n$. Let $A \subseteq U \times \mathbb{R}^n$ be Jordan-measurable, and let $t \in \mathbb{R}$. Since ϕ^t is smooth and injective, the Change of Variables Theorem implies that $\phi^t(A)$ is Jordan-measurable, and

$$\mathrm{vol}(\phi^t(A)) = \int_A |\det(D_x\phi(t,\cdot))|.$$

If F is the Hamiltonian vector field associated to ϕ^t, then F is differentiable, and

$$\mathrm{tr}(DF) = \mathrm{div}\, F.$$

By Hamilton's equations,

$$\mathrm{div}\, F = \frac{\partial}{\partial q}\left(-\frac{\partial H}{\partial p}\right) + \frac{\partial}{\partial p}\left(\frac{\partial H}{\partial q}\right) = 0.$$

According to the First Variation Equation, every matrix representation of $D_x\phi(t,x)$ is a fundamental matrix solution of the linear differential equation

$$\dot{y} = DF_{\phi^t(x)}y,$$

and

$$\det(D_x\phi(0,x)) = 1 \neq 0.$$

Thus, the Liouville-Ostrogradski Formula (Proposition 4.2.8) implies that

$$\begin{aligned}
\mathrm{vol}(\phi^t(A)) &= \int_A |\det(D_x\phi(t,\cdot))| \\
&= \int_A \exp\left(\int_0^t \mathrm{tr}(DF(\phi(s,\cdot)))\, ds\right) \\
&= \int_A \exp\left(\int_0^t (\mathrm{div}\ F)(\phi(s,\cdot))\, ds\right) \\
&= \int_A \exp(0) \\
&= \int_A 1 \\
&= \int_{\mathbb{R}^n\times\mathbb{R}^n} \chi_A \\
&= \mathrm{vol}\ (A).
\end{aligned}$$

Therefore, ϕ^t is volume-preserving. □

Volume preservation has significant consequences for recurrence.

Theorem 4.2.17. The Poincaré Recurrence Theorem. *If ϕ^t is a volume-preserving flow on a bounded subset D of $\mathbb{R}^n$, then*

$$\Omega(\phi^t) = D.$$

Proof. Let U be a nonempty open subset of D. Every nonempty open set contains a nonempty Jordan-measurable subset. For instance, every nonempty open set contains a nonempty open ball. Thus, without loss of generality, we may assume that U is Jordan-measurable.

Since ϕ^t is volume-preserving, the sets $\phi^1(U)$, $\phi^2(U)$, $\phi^3(U), \ldots$ are Jordan-measurable and have equal volumes. Consequently,

$$\sum_{n=1}^{\infty} \mathrm{vol}(\phi^n(U))$$

is infinite.

By means of contradiction assume that the sets $\phi^1(U)$, $\phi^2(U)$, $\phi^3(U), \ldots$ are pairwise disjoint. In this case,

$$\sum_{n=1}^{\infty} \mathrm{vol}(\phi^n(U)) = \mathrm{vol}\left(\bigcup_{n=1}^{\infty} \phi^n(U)\right)$$

where we consider these quantities as extended real numbers. Since

$$\bigcup_{n=1}^{\infty} \phi^n(U) \subseteq D$$

and D is bounded we obtain

$$\sum_{n=1}^{\infty} \operatorname{vol}(\phi^n(U)) = \operatorname{vol}\left(\bigcup_{n=1}^{\infty} \phi^n(U)\right) \leq \operatorname{vol}(D) < \infty.$$

This is a contradiction. So, the sets $\phi^1(U)$, $\phi^2(U)$, $\phi^3(U), \ldots$ are not pairwise disjoint. Hence, there exist integers $l \geq 1$ and $k > l$ such that

$$\phi^k(U) \cap \phi^l(U) \neq \emptyset.$$

Thus, if $T = k - l$, then $T \geq 1$, and

$$\phi^T(U) \cap U \neq \emptyset.$$

Therefore,

$$\Omega(\phi^t) = D.$$

□

The Poincaré Recurrence Theorem originally appears in Poincaré [**34**]. The study of volume-preserving flows motivates the study of ergodic theory. See Walters [**44**].

4.3. Invariant Functions

Corollary 4.2.6 showed that a Hamiltonian function is constant on each orbit of its corresponding Hamiltonian flow. A real-valued function which is constant on each orbit of a flow is called an invariant function.

Definition 4.3.1. Let ϕ^t be a flow on a topological space X. A function $f : X \to \mathbb{R}$ is **invariant** with respect to ϕ^t if $f(\phi^t(x)) = f(x)$ for all $t \in \mathbb{R}$ and all $x \in X$.

Example 4.3.2. Every Hamiltonian function is an invariant function with respect to its corresponding Hamiltonian flow by Corollary 4.2.6.

Continuous invariant functions are constant on topologically transitive sets.

Proposition 4.3.3. *Let ϕ^t be a flow on a topological space X. If X is topologically transitive with respect to ϕ^t, then every continuous invariant function with respect to ϕ^t is constant.*

Proof. Let $f : X \to \mathbb{R}$ be continuous and invariant with respect to ϕ^t. Since X is topologically transitive with respect to ϕ^t, there exists $x \in X$ such that $X = \operatorname{cl}(\mathcal{O}^+(x))$. The invariance of f with respect to ϕ^t implies that there exists a real number c such that $f(\mathcal{O}^+(x)) = \{c\}$. Since $f(X) = f(\operatorname{cl}(\mathcal{O}^+(x))) \subseteq \operatorname{cl}(f(\mathcal{O}^+(x))) = \{c\}$, the function f is constant. □

4.4. Weak and Strong Lyapunov Functions

Returning to the harmonic oscillator of Example 4.2.3 and the ideal pendulum of Example 4.2.4, we examine how damping affects total energy and develop two types of test functions named for the Russian mathematician Alexander M. Lyapunov (1857–1918).

Example 4.4.1. Consider the flow on $\mathbb{R} \times \mathbb{R}$ associated to the system of differential equations

$$\begin{aligned} \dot{x} &= y, \\ \dot{y} &= -x - \delta y \end{aligned}$$

modeling a **damped harmonic oscillator** with displacement x, velocity y and damping factor $\delta > 0$. As in Example 4.2.3 the total energy function of this oscillator is the function $L : \mathbb{R} \times \mathbb{R} \to \mathbb{R}$ defined by

$$L(x, y) = \frac{1}{2}(x^2 + y^2).$$

Computing the derivative of L along a solution $(x(t), y(t))$ we obtain

$$\begin{aligned} \frac{d}{dt} L(x(t), y(t)) &= x(t)\dot{x}(t) + y(t)\dot{y}(t) \\ &= x(t)y(t) - y(t)[x(t) + \delta y(t)] \\ &= -\delta [y(t)]^2 \\ &\leq 0 \end{aligned}$$

with equality if and only if $y(t) = 0$. Computing higher derivatives of L along $(x(t), y(t))$ we find

$$\begin{aligned} \frac{d^2}{dt^2} L(x(t), y(t)) &= -2\delta y(t)\dot{y}(t) \\ &= 2\delta y(t)[x(t) + \delta y(t)], \end{aligned}$$

and

$$\begin{aligned} \frac{d^3}{dt^3} L(x(t), y(t)) &= 2\delta\big(\dot{y}(t)[x(t) + \delta y(t)] + y(t)[\dot{x}(t) + \delta \dot{y}(t)]\big) \\ &= 2\delta\big((1 - 2\delta^2)[y(t)]^2 - 3\delta y(t)x(t) - [x(t)]^2\big). \end{aligned}$$

If $x(t_0) \neq 0$ and $y(t_0) = 0$, then

$$\left.\frac{d^2}{dt^2}\right|_{t=t_0} L(x(t), y(t)) = 0,$$

and

$$\left.\frac{d^3}{dt^3}\right|_{t=t_0} L(x(t), y(t)) = -2\delta [x(t_0)]^2 < 0.$$

If $x(t_0) = 0$ and $y(t_0) = 0$, then the solution is fixed. Therefore, L decreases along orbits.

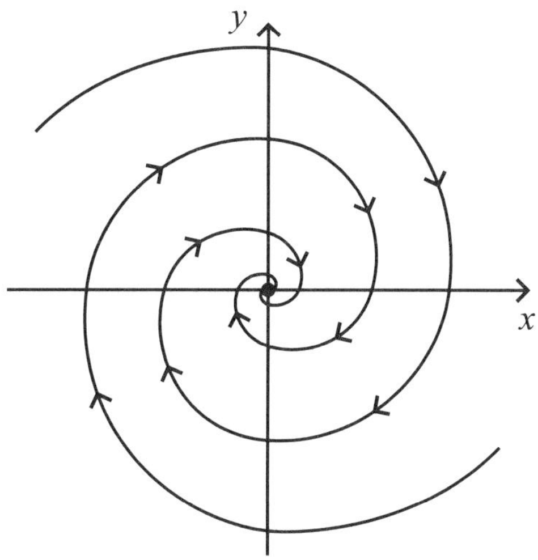

Figure 4.11. A phase portrait of a damped harmonic oscillator

However,

$$\frac{\partial}{\partial y} y = 1 \neq -1 = \frac{\partial}{\partial x}(-x - \delta y).$$

So, this flow is not a gradient flow by Proposition 4.1.5. See Figure 4.11.

Example 4.4.2. Consider the flow on $\mathbb{R} \times \mathbb{R}$ associated to the system of differential equations

$$\begin{aligned} \dot{x} &= y, \\ \dot{y} &= -\sin x - \delta y \end{aligned}$$

describing the motion of a **damped pendulum** with angular displacement x from a downward equilibrium, angular velocity y and damping factor $\delta > 0$. As in Example 4.2.4 the total energy function for the pendulum is the function $L : \mathbb{R} \times \mathbb{R} \to \mathbb{R}$ defined by

$$L(x, y) = \frac{1}{2}y^2 + 1 - \cos x.$$

Computing the derivative of L along a solution $(x(t), y(t))$ we obtain

$$\begin{aligned} \frac{d}{dt}L(x(t), y(t)) &= y(t)\dot{y}(t) + \sin(x(t))\dot{x}(t) \\ &= y(t)[-\sin(x(t)) - \delta y(t)] + \sin(x(t))y(t) \\ &= -\delta[y(t)]^2 \\ &\leq 0 \end{aligned}$$

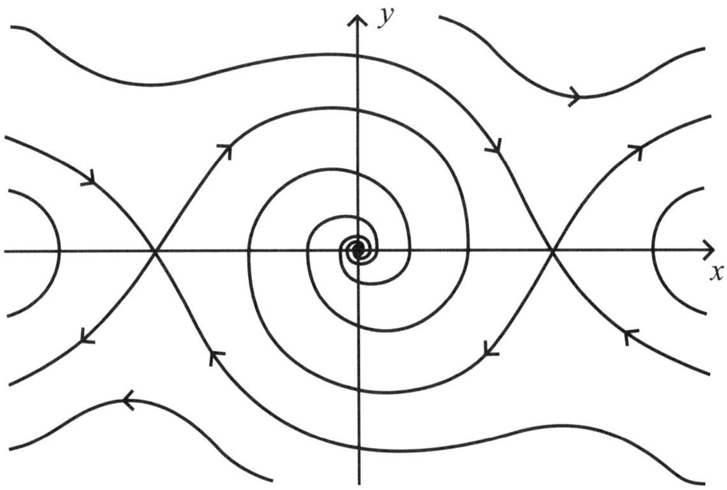

Figure 4.12. A phase portrait of a damped pendulum

with equality if and only if $y(t) = 0$. Computing higher derivatives of L along $(x(t), y(t))$ we find

$$\begin{aligned}\frac{d^2}{dt^2}L(x(t),y(t)) &= -2\delta y(t)\dot{y}(t)\\ &= 2\delta y(t)\big(\sin[x(t)] + \delta y(t)\big),\end{aligned}$$

and

$$\begin{aligned}\frac{d^3}{dt^3}L(x(t),y(t)) &= 2\delta\big[\dot{y}(t)\big(\sin[x(t)] + \delta y(t)\big) + y(t)\big(\cos[x(t)]\dot{x}(t) + \delta\dot{y}(t)\big)\big]\\ &= 2\delta\Big[\big(\cos[x(t)] - 2\delta^2\big)[y(t)]^2 - 3\delta y(t)\sin[x(t)] - \sin^2[x(t)]\Big].\end{aligned}$$

If $x(t_0)$ is not an integral multiple of π and $y(t_0) = 0$, then

$$\left.\frac{d^2}{dt^2}\right|_{t=t_0} L(x(t),y(t)) = 0,$$

and

$$\left.\frac{d^3}{dt^3}\right|_{t=t_0} L(x(t),y(t)) = -2\delta\sin^2[x(t_0)] < 0.$$

If $x(t_0)$ is an integral multiple of π and $y(t_0) = 0$, then the solution is fixed. Therefore, L decreases along orbits.

However,

$$\frac{\partial}{\partial y}y = 1 \neq -\cos x = \frac{\partial}{\partial x}(-\sin x - \delta y).$$

So, this flow is not a gradient flow by Proposition 4.1.5. See Figure 4.12.

For the damped harmonic oscillator in Example 4.4.1 and the damped pendulum in Example 4.4.2 total energy decreases along orbits. Thus, total energy is not a Hamiltonian function for these systems. Instead, total energy decreases along orbits like a potential function. Yet, neither of these flows is a gradient flow. The total energy functions in Examples 4.4.1 and 4.4.2 are examples of weak Lyapunov functions.

Definition 4.4.3. Let ϕ^t be a flow on a topological space X. A continuous function $L : X \to \mathbb{R}$ is a **weak Lyapunov function** with respect to ϕ^t if $L(\phi^t(x)) \leq L(x)$ for all $x \in X$ and all $t > 0$.

Example 4.4.4. Every continuous invariant function is a weak Lyapunov function. In particular, every Hamiltonian function is a weak Lyapunov function.

Example 4.4.5. Every potential function is a weak Lyapunov function by Proposition 4.1.11.

Proposition 4.1.11 showed that a potential function strictly decreases on the complement of the set of fixed points of the corresponding gradient flow. Now we consider flows that admit weak Lyapunov functions which strictly decrease on the complement of the set of fixed points.

Definition 4.4.6. Let ϕ^t be a flow on a topological space X. A function $L : X \to \mathbb{R}$ is a **strong Lyapunov function** with respect to ϕ^t if L is a weak Lyapunov function with respect to ϕ^t and $L(\phi^t(x)) < L(x)$ for all $t > 0$ and all $x \in X \setminus \mathrm{Fix}(\phi^t)$.

Example 4.4.7. Every potential function is a strong Lyapunov function by Proposition 4.1.11.

Definition 4.4.8. A flow ϕ^t on a topological space X is **gradient-like** if there exists a strong Lyapunov function $L : X \to \mathbb{R}$ with respect to ϕ^t.

Example 4.4.9. Every gradient flow is gradient-like because every potential function is a strong Lyapunov function.

The following example shows that not every gradient-like flow is a gradient flow.

Example 4.4.10. Consider the flow ϕ^t of the system of differential equations

$$\begin{aligned} \dot{x} &= -x + y, \\ \dot{y} &= -x - y \end{aligned}$$

on $\mathbb{R}^2$.

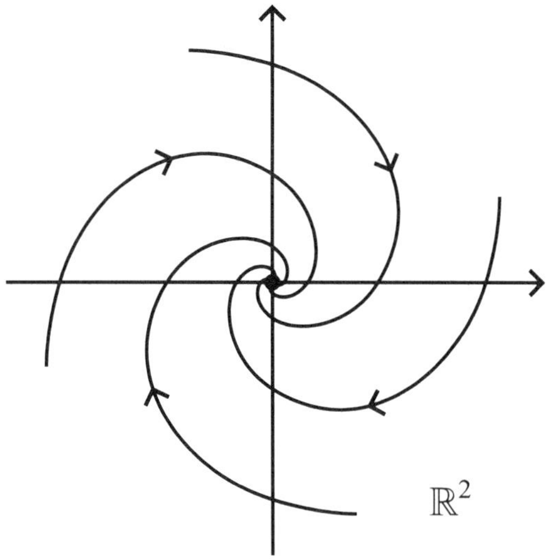

Figure 4.13. A phase portrait of the flow in Example 4.4.10

Define $L : \mathbb{R}^2 \to \mathbb{R}$ by

$$L(x, y) = x^2 + y^2.$$

Computing the derivative of L along an orbit we obtain

$$\begin{aligned}\frac{d}{dt}L(x(t), y(t)) &= 2x(t)\dot{x}(t) + 2y(t)\dot{y}(t) \\ &= 2x(t)[-x(t) + y(t)] + 2y(t)[-x(t) - y(t)] \\ &= -2\big([x(t)]^2 + [y(t)]^2\big) \\ &\leq 0\end{aligned}$$

with equality if and only if $(x, y) = (0, 0)$. Since $(0, 0)$ is the unique fixed point of ϕ^t, the function L is a strong Lyapunov function with respect to ϕ^t. Therefore, ϕ^t is a gradient-like flow. However,

$$\frac{\partial}{\partial y}(-x + y) = 1 \neq -1 = \frac{\partial}{\partial x}(-x - y).$$

Therefore, ϕ^t is not a gradient flow by Proposition 4.1.5. See Figure 4.13.

Example 4.4.11. Consider the flow ψ^t associated with the system of differential equations

$$\begin{aligned}\dot{\theta} &= \sin^2\theta + 1 - \left(\frac{\phi}{\pi}\right)^2, \\ \dot{\phi} &= \phi\left(1 - \frac{\phi}{\pi}\right)\left[\sin^2\theta + 1 - \left(\frac{\phi}{\pi}\right)^2\right]\end{aligned}$$

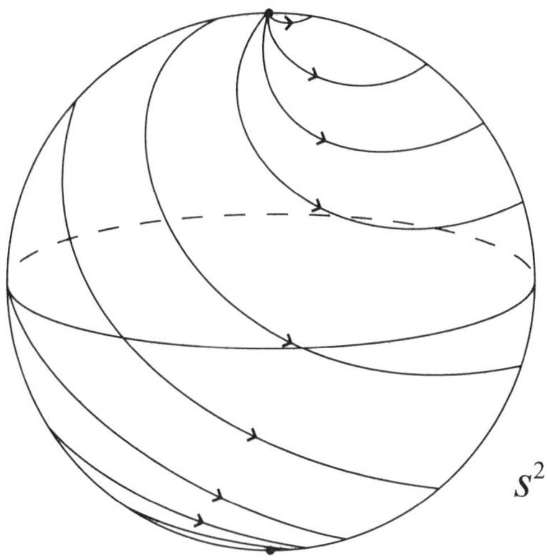

Figure 4.14. A phase portrait of the flow in Example 4.4.11

in spherical coordinates on S^2. In spherical coordinates the fixed points of ψ^t are $(0,0)$ and $(0,\pi)$. See Figure 4.14.

Define $L : S^2 \to \mathbb{R}$ by

$$L(\theta,\phi) = 1 - \frac{\phi}{\pi}.$$

Computing the derivative of L along a solution $(\theta(t),\phi(t))$ we obtain

$$\begin{aligned} \frac{d}{dt}L(\theta(t),\phi(t)) &= -\frac{\dot{\phi}(t)}{\pi} \\ &= -\frac{\phi(t)}{\pi}\left[1-\frac{\phi(t)}{\pi}\right]\left(\sin^2[\theta(t)] + 1 - \left[\frac{\phi(t)}{\pi}\right]^2\right) \\ &\leq 0 \end{aligned}$$

with equality if and only if $\phi = 0$ or $\phi = \pi$. Thus, $L(\psi^t(\theta,\phi)) \leq L(\theta,\phi)$ for all $(\theta,\phi) \in S^2$ and all $t > 0$ with equality if and only if $(\theta,\phi) \in \text{Fix}(\psi^t)$. Therefore, L is a strong Lyapunov function with respect to ψ^t, and ψ^t is a gradient-like flow.

To further generalize gradient flows, we revisit them in a more abstract setting.

4.5. Potential Functions on Riemannian Manifolds

Section 4.1 defined gradient flows on open subsets of Euclidean spaces and on surfaces in $\mathbb{R}^3$. Recall that the gradient of a differentiable real-valued function in a neighborhood of a surface in $\mathbb{R}^3$ is a vector field on that neighborhood. By projecting this vector field into the tangent planes to the surface we obtain a vector field on the surface which defines a gradient flow via a differential equation.

The objective of this section is to define gradient flows on Riemannian manifolds. At each point p on a differentiable manifold M the derivative at p of a differentiable real-valued function on M is a linear functional from the tangent space to M at p to the set of real numbers. The idea is to identify this linear functional with an element of the tangent space at p. Carrying out this operation at each point of M defines a vector field on M generating a flow on M via a differential equation. We now describe this process in detail.

A Riemannian metric g on a manifold M associates to each $p \in M$ a symmetric positive-definite bilinear real-valued function g_p on the tangent space TM_p which varies smoothly with p. The standard inner product on $\mathbb{R}^n$ is an example of a Riemannian metric. The pair (M, g) is called a Riemannian manifold. See Hicks [**20**] for a complete development of Riemannian metrics.

Let (M, g) be a Riemannian manifold. For each $p \in M$ and $v \in TM_p$, define a function $L_{p,v} : TM_p \to \mathbb{R}$ by

$$L_{p,v}(w) = g_p(v, w).$$

Since g_p is bilinear, the function $L_{p,v}$ is a linear transformation. The dual space of a finite-dimensional real linear space V is the set V^* of linear transformations mapping V to $\mathbb{R}$. Thus, $L_{p,v} \in (TM_p)^*$. For each $p \in M$ define $J_p : TM_p \to (TM_p)^*$ by

$$J_p(v) = L_{p,v}.$$

Since g_p is positive definite for each $p \in M$, the map J_p is a linear space isomorphism for each $p \in M$.

Let $V : M \to \mathbb{R}$ be a differentiable function. For each $p \in M$ the derivative $DV_p : TM_p \to \mathbb{R}$ is an element of $(TM_p)^*$.

Definition 4.5.1. Let M be a Riemannian manifold. Let $V : M \to \mathbb{R}$ be differentiable. Define the **gradient vector** of V at $p \in M$ by

$$\operatorname{grad}(V)_p = J_p^{-1}(DV_p).$$

Definition 4.5.2. A vector field F on a Riemannian manifold M is a **gradient vector field** if there exists a differentiable function $V : M \to \mathbb{R}$ such that

$$F(p) = -\operatorname{grad}(V)_p$$

for all $p \in M$. A differentiable flow ϕ^t on M is a **gradient flow** if there exists a gradient vector field F such that

$$\left.\frac{d}{dt}\right|_{t=0} \phi^t(p) = F(p)$$

for all $p \in M$, in which case there exists a differentiable function $V : M \to \mathbb{R}$ such that

$$\left.\frac{d}{dt}\right|_{t=0} \phi^t(p) = -\text{grad}(V)_p$$

for all $p \in M$. The function V is a **potential function** for the vector field F or for the flow ϕ^t.

The fixed points of a gradient flow ϕ^t with potential function V are precisely the zeros of grad(V). A gradient flow on a Riemannian manifold is gradient-like, and every potential function is a strong Lyapunov function with respect to its gradient flow.

Proposition 4.5.3. *Let ϕ^t be a gradient flow on a Riemannian manifold M. If V is a potential function for ϕ^t, then $V(\phi^t(x)) \leq V(x)$ for each $x \in M$ and for all $t \geq 0$ with equality if and only if $x \in Fix(\phi^t)$.*

Proof. Let $x \in M$. By the Chain Rule,

$$\begin{aligned}
\left.\frac{d}{dt}\right|_{t=0} V(\phi^t(x)) &= DV_{\phi^0(x)}\left(\left.\frac{d}{dt}\right|_{t=0} \phi^t(x)\right) \\
&= DV_x(-\text{grad}(V)_x) \\
&= -\|\text{grad}(V)_x\|^2 \\
&\leq 0
\end{aligned}$$

with equality if and only if grad$(V)_x$ is zero. Therefore, $V(\phi^t(x)) \leq V(x)$ for all $x \in M$ and $t \geq 0$ with equality if and only if $x \in \text{Fix}(\phi^t)$. □

Since a Riemannian manifold is a metric space, we can investigate the chain recurrent sets of flows on Riemannian manifolds. Proposition 4.5.5 will show that if ϕ^t is a gradient flow on a compact Riemannian manifold and the image of Fix(ϕ^t) under a potential function is finite, then $\mathcal{R}(\phi^t) = \text{Fix}(\phi^t)$. In that direction we prove a lemma. Compare the proofs of Lemma 4.5.4 and Proposition 4.5.5 with Examples 2.7.5 and 2.7.9.

Lemma 4.5.4. *Let ϕ^t be a gradient flow with a potential function V on a compact Riemannian manifold M. If $c \notin V(Fix(\phi^t))$, then for each $T > 0$ there exists $\delta > 0$ such that if $y \in M$ and $V(y) < c + \delta$, then $V(\phi^t(y)) < c$ for all $t \geq T$.*

Proof. Let T be a positive real number. By means of contradiction assume that there exist sequences $\{y_k\}$ in M and $\{s_k\}$ in $[T, \infty)$ such that for each positive integer k

$$V(y_k) < c + \frac{1}{k},$$

and

$$V(\phi^{s_k}(y_k)) \geq c.$$

Since V decreases along orbits,

$$c \leq V(\phi^{s_k}(y_k)) \leq V(\phi^T(y_k)) \leq V(y_k) < c + \frac{1}{k}$$

for each positive integer k. The compactness of M implies that the sequence $\{y_k\}$ has a limit point $\bar{y} \in M$. By the continuity of V and ϕ^T,

$$c = V\left(\phi^T(\bar{y})\right) = V(\bar{y}).$$

However, V is strictly decreasing along orbits in $M \backslash \text{Fix}(\phi^t)$. So, $\bar{y} \in \text{Fix}(\phi^t)$, contradicting the fact that

$$V(\bar{y}) = c \notin V\left(\text{Fix}(\phi^t)\right).$$

Therefore, for each $T > 0$ there exists $\delta > 0$ such that if $y \in M$ and $V(y) < c + \delta$, then $V\left(\phi^t(y)\right) < c$ for all $t \geq T$. □

Proposition 4.5.5. *If ϕ^t is a gradient flow with a potential function V on a compact Riemannian manifold M and $V(Fix(\phi^t))$ is finite, then*

$$\mathcal{R}(\phi^t) = Fix(\phi^t).$$

Remark 4.5.6. In fact, Proposition 4.5.5 is true without the hypothesis that $V(\text{Fix}(\phi^t))$ is finite. We leave this more general result to Exercise 15 for consideration after reading Section 4.8.

Proof. Example 2.7.5 shows that $\text{Fix}(\phi^t) \subseteq \mathcal{R}(\phi^t)$. It remains to prove the reverse inclusion. We will demonstrate that if $x \in M \setminus \text{Fix}(\phi^t)$, then $x \notin \mathcal{R}(\phi^t)$ by showing that for each $T > 0$ there exists an $\epsilon > 0$ such that there does not exist an (ϵ, T)-chain from x to itself.

Let $x \in M \setminus \text{Fix}(\phi^t)$, and let $T > 0$. By Proposition 4.5.3 the function $V(\phi^t(x))$ is strictly decreasing with respect to t. Since $V(\text{Fix}(\phi^t))$ is finite, there is a positive real number S such that

$$V(\phi^t(x)) \notin V(\text{Fix}(\phi^t))$$

for all $t \geq S$.

By Corollary 3.3.8 the chain recurrent set of ϕ^t is invariant. So, $x \notin \mathcal{R}(\phi^t)$ if and only if $\phi^S(x) \notin \mathcal{R}(\phi^t)$. Hence, without loss of generality, we may assume that

$$V(\phi^t(x)) \notin V(\text{Fix}(\phi^t))$$

for all $t \geq 0$.

If $b = V(x)$ and $a = V(\phi^T(x))$, then $a < b$. By Lemma 4.5.4 there exists $\delta > 0$ such that if $V(y) < a + \delta$, then

$$V(\phi^t(y)) < a$$

for all $t \geq T$.

Let $\eta = \min\{\delta, \frac{1}{2}(b-a)\}$, and let d denote the metric on M induced by the Riemannian metric on M. Since M is compact, V is uniformly continuous. So, there exists $\epsilon > 0$ such that if $y_1, y_2 \in M$ and $d(y_1, y_2) < \epsilon$, then

$$|V(y_1) - V(y_2)| < \eta.$$

By means of contradiction assume that $x \in \mathcal{R}(\phi^t)$. Then there is an (ϵ, T)-chain

$$(x = x_0, x_1, \ldots, x_n = x; t_0, \ldots t_{n-1})$$

from x to itself. Since $d(\phi^{t_0}(x_0), x_1) < \epsilon$, the uniform continuity of V implies that

$$|V(\phi^{t_0}(x_0)) - V(x_1)| < \eta \leq \delta.$$

Because $t_0 \geq T$,

$$V(x_1) < V(\phi^{t_0}(x_0)) + \delta \leq V(\phi^T(x_0)) + \delta = a + \delta.$$

Since $t_1 \geq T$,

$$V(\phi^{t_1}(x_1)) < a$$

by the choice of δ.

Since $d(\phi^{t_1}(x_1), x_2) < \epsilon$, the uniform continuity of V implies that

$$|V(\phi^{t_1}(x_1)) - V(x_2)| < \eta \leq \delta.$$

Hence,

$$V(x_2) < V(\phi^{t_1}(x_1)) + \delta \leq a + \delta.$$

Because $t_2 \geq T$,

$$V(\phi^{t_2}(x_2)) < a$$

by the choice of δ. Continuing in this manner,

$$V(\phi^{t_{n-1}}(x_{n-1})) < a.$$

Since $V(x_n) = V(x) = b > a$,

$$|V(\phi^{t_{n-1}}(x_{n-1})) - V(x_n)| > b - a\,.$$

However, because $d(\phi^{t_{n-1}}(x_{n-1}), x_n) < \epsilon$,

$$|V(\phi^{t_{n-1}}(x_{n-1})) - V(x_n)| < \eta \leq \frac{1}{2}(b-a)\,.$$

This is a contradiction. Consequently, there is no (ϵ, T)-chain from x to itself. Therefore,

$$\mathcal{R}(\phi^t) = \mathrm{Fix}(\phi^t).$$

□

4.6. Morse Functions

During the 1920s Marston Morse (1892–1977) introduced the study of a manifold M via certain smooth real-valued functions defined on M now called Morse functions. See Milnor [**30**] for a treatment of Morse Theory. In this section we investigate potential functions which are also Morse functions. All manifolds under consideration have positive dimension.

Definition 4.6.1. Let M be a manifold, and let $f : M \to \mathbb{R}$ be a differentiable function. A point $p \in M$ is a **regular point** of f if Df_p is surjective. A point $p \in M$ is a **critical point** of f if p is not a regular point of f.

A point p is a critical point of a differentiable function $f : M \to \mathbb{R}$ whenever $Df_p(v) = 0$ for all $v \in TM_p$. If $(x_1, ..., x_n)$ is a local coordinate system on a neighborhood of p in M, then p is a critical point of f if and only if

$$\frac{\partial f}{\partial x_1}(p) = \cdots = \frac{\partial f}{\partial x_n}(p) = 0.$$

If ϕ^t is a gradient flow with a potential function V, then the fixed points of ϕ^t are exactly the zeros of $\mathrm{grad}(V)$, which are exactly the critical points of V.

Definition 4.6.2. Let V and W be linear spaces. A bilinear map $B : V \times W \to \mathbb{R}$ is **nondegenerate** if $B(v, w) = 0$ for all $w \in W$ implies that $v = 0$.

Definition 4.6.3. Let M be a manifold. If $f : M \to \mathbb{R}$ is twice differentiable, then a critical point p of f is called **nondegenerate** if the bilinear map $D^2 f_p : TM_p \times TM_p \to \mathbb{R}$ is nondegenerate.

If $(x_1, ..., x_n)$ is a local coordinate system for a neighborhood in M of a critical point p of f, then p is a nondegenerate critical point of f if and only if the Hessian matrix

$$\left(\frac{\partial^2 f}{\partial x_i \partial x_j}(p) \right)$$

is nonsingular. Proposition 4.6.4 shows that the nondegenerate critical points of a C^2 (twice continuously differentiable) real-valued function on a manifold are isolated.

Proposition 4.6.4. *Let f be a C^2 real-valued function on a manifold. If p is a nondegenerate critical point of f, then there exists an open neighborhood of p in which p is the only critical point of f.*

Proof. Let p be a nondegenerate critical point of f, and let $(x_1, ..., x_n)$ be a local coordinate system on a neighborhood U of p. Define $g : U \to \mathbb{R}^n$ by

$$g(x) = \left(\frac{\partial f}{\partial x_1}(x), ..., \frac{\partial f}{\partial x_n}(x)\right).$$

Since p is a critical point, $g(p) = (0, 0, \ldots, 0)$. The matrix of Dg_p in the coordinates $(x_1, \ldots, x_n)$ is

$$\left(\frac{\partial^2 f}{\partial x_i \partial x_j}(p)\right),$$

which is nonsingular since p is nondegenerate. By the Inverse Function Theorem, there exists an open neighborhood W of p such that g maps W diffeomorphically onto a neighborhood of $(0, 0, \ldots, 0)$. Thus, g cannot be zero at any other point in W. Therefore, p is the only critical point of f in W. □

Definition 4.6.5. Let M be a manifold. A C^2 function $f : M \to \mathbb{R}$ is a **Morse function** if each of its critical points is nondegenerate.

Example 4.6.6. The height function $f : S^2 \to \mathbb{R}$ of Example 4.1.9 defined by $f(x, y, z) = z$ is a Morse function. We leave the verification of the details to Exercise 11.

Example 4.6.7. The height function $f : \mathbb{T}^2 \to \mathbb{R}$ of Example 4.1.10 defined by $f(x, y, z) = z$ is a Morse function. We leave the verification of the details to Exercise 12.

By Proposition 4.6.4 the critical points of a Morse function on a manifold are isolated. The following corollary establishes that if the manifold is compact, then there are at most finitely many critical points.

Corollary 4.6.8. *A Morse function on a compact manifold has finitely many critical points.*

Proof. Let f be a Morse function on a compact manifold M. Let C denote the set of critical points of f. We claim that $M \setminus C$ is open in M. If $x \in M \setminus C$, then $Df_x \neq 0$. Since f is C^1 there exists an open neighborhood U of x such that $Df_y \neq 0$ for all $y \in U$. So, $U \subseteq M \setminus C$. Thus, $M \setminus C$ is open in M.

By Proposition 4.6.4 for each $p \in C$ there exists an open neighborhood W_p of p in M such that $W_p \cap C = \{p\}$. Define $\mathcal{C} = \{W_p \mid p \in C\} \cup \{M \setminus C\}$. The collection $\mathcal{C}$ is an open cover of M, and each element of $\mathcal{C}$ contains at most one critical point of f. Since M is compact, the collection $\mathcal{C}$ has a finite subcover of M. Therefore, f has finitely many critical points. □

In fact, a typical Morse function on a compact manifold assumes distinct values at distinct critical points.

Proposition 4.6.9. *Let M be a compact manifold. If f is a Morse function on M, then for every $\epsilon > 0$ there exists a Morse function g on M such that g is injective on its set of critical points, and*

$$\sup_{x \in M} |g(x) - f(x)| < \epsilon.$$

Proof. Let f be a Morse function on M, and let $\epsilon > 0$. By Corollary 4.6.8, f has finitely many critical points $p_1, \ldots, p_n$. There exist disjoint neighborhoods $U_1, \ldots, U_n$ such that $p_i \in U_i$ for $i = 1, \ldots, n$. Let $V_1, \ldots, V_n$ be neighborhoods of $p_1, \ldots, p_n$, respectively, such that $\mathrm{cl}(V_i)$ is a proper subset of U_i for $i = 1, \ldots, n$. There exist C^2 real-valued functions $\rho_1, \ldots, \rho_n$ (called bump functions) such that $\rho_i|_{V_i} = 1$ and $\rho_i|_{M \setminus U_i} = 0$ for $i = 1, \ldots, n$. Since M is compact, define

$$\delta = \min\left\{ \|Df_x\| \mid x \in \mathrm{cl}\left(\bigcup_{i=1}^{n} (U_i \setminus V_i) \right) \right\},$$

and

$$K = \max\{\|(D\rho_i)_x\| \mid i = 1, \ldots, n \text{ and } x \in M\}.$$

The fact that f has no critical points in the compact set $\mathrm{cl}\big(\bigcup_{i=1}^{n}(U_i \setminus V_i)\big)$ implies that $\delta > 0$. Since each of $D\rho_1, \ldots, D\rho_n$ is nonsingular at some point, $K > 0$.

There exist real numbers $a_1, \ldots, a_n$ such that

$$f(p_i) + a_i \neq f(p_j) + a_j$$

whenever $i \neq j$, and

$$\sum_{i=1}^{n} |a_i| < \min\left\{ \epsilon, \frac{\delta}{K} \right\}.$$

Define $g : M \to \mathbb{R}$ by

$$g(x) = f(x) + \sum_{i=1}^{n} a_i \rho_i(x).$$

The function g is C^2 since it is a linear combination of C^2 functions, and

$$Dg_x = Df_x + \sum_{i=1}^{n} a_i (D\rho_i)_x$$

for each $x \in M$. Since $\rho_1, \ldots, \rho_n$ are constant on $V_1, \ldots, V_n$,

$$(Dg)|_{V_i} = (Df)|_{V_i}$$

for $i = 1, \ldots, n$. Consequently,

$$(D^2 g)|_{V_i} = (D^2 f)|_{V_i}$$

for $i = 1, \ldots, n$.

If x is a critical point of f, then $Df_x = Dg_x$. So, x is a critical point of g. Since f is a Morse function, x is a nondegenerate critical point of f. Because $D^2 g_x = D^2 f_x$, the point x is a nondegenerate critical point of g as well.

Let x be a regular point of f. If $x \in M \setminus \mathrm{cl}\big(\bigcup_{i=1}^n (U_i \setminus V_i)\big)$, then $Dg_x = Df_x$, so that x is a regular point of g. If $x \in \mathrm{cl}\big(\bigcup_{i=1}^n (U_i \setminus V_i)\big)$, then $\|Df_x\| > \delta > 0$. Since

$$\|Dg_x - Df_x\| = \left\| \sum_{i=1}^n a_i (D\rho_i)_x \right\| \le K \sum_{i=1}^n |a_i| < \delta,$$

we obtain

$$\|Dg_x\| > 0.$$

Thus, x is a regular point of g. Therefore, f and g have the same critical points, and g is a Morse function on M. If $g(p_i) = g(p_j)$, then

$$f(p_i) + a_i = f(p_j) + a_j.$$

By the selection of $a_1, \ldots, a_n$ we have $p_i = p_j$. Thus, g is injective on its set of critical points.

Finally,

$$\begin{aligned} |g(x) - f(x)| &= |f(x) + \sum_{i=1}^n a_i \rho_i(x) - f(x)| \\ &\le \sum_{i=1}^n |a_i| \\ &< \epsilon. \end{aligned}$$

Therefore,

$$\sup_{x \in M} |g(x) - f(x)| < \epsilon. \qquad \square$$

Stated abstractly, Proposition 4.6.9 says that the set of Morse functions on a compact manifold M which assume distinct values at distinct critical points is dense in the set of all Morse functions on M with respect to the topology of uniform convergence on the set of Morse functions.

Now consider the gradient flow of a Morse function. Proposition 4.6.9 guarantees that a typical Morse function on a compact manifold is injective on the set of fixed points of its gradient flow.

Corollary 4.6.10. *If ϕ^t is the gradient flow of a Morse function f on a compact Riemannian manifold, then Fix(ϕ^t) and f(Fix(ϕ^t)) are finite.*

Proof. The fixed points of ϕ^t are exactly the critical points of f. By Corollary 4.6.8, f has finitely many critical points. So, $\mathrm{Fix}(\phi^t)$ is finite. The image of any finite set is finite. Thus, $f(\mathrm{Fix}(\phi^t))$ is finite. □

Proposition 4.6.11. *If ϕ^t is the gradient flow of a Morse function f on a compact Riemannian manifold, then*

$$\mathcal{R}(\phi^t) = \mathit{Fix}(\phi^t).$$

Proof. By Corollary 4.6.10 the set $f(\mathrm{Fix}(\phi^t))$ is finite. Applying Proposition 4.5.5 we obtain $\mathcal{R}(\phi^t) = \mathrm{Fix}(\phi^t)$. □

Example 4.6.12. Let ϕ^t be the height flow on $\mathbb{T}^2$. Example 4.1.15 shows that $\mathcal{R}(\phi^t) = \mathrm{Fix}(\phi^t)$.

Because the fixed points of the gradient flow of a Morse function are isolated, they form a totally disconnected set. Thus, the chain recurrent set of the gradient flow of a Morse function on a compact Riemannian manifold is totally disconnected.

Definition 4.6.13. A flow on a metric space is called **strongly gradient-like** if its chain recurrent set is totally disconnected.

Proposition 4.6.14. *If ϕ^t is a strongly gradient-like flow, then*

$$\mathcal{R}(\phi^t) = \mathit{Fix}(\phi^t).$$

Proof. By Example 2.7.5 it suffices to prove that $\mathcal{R}(\phi^t) \subseteq \mathrm{Fix}(\phi^t)$. Let $x \in \mathcal{R}(\phi^t)$. By Corollary 3.3.8 the set $\mathcal{R}(\phi^t)$ is invariant. Therefore, $\mathcal{O}(x) \subseteq \mathcal{R}(\phi^t)$. By Proposition 1.1.11 the orbit $\mathcal{O}(x)$ is connected. So, a single connected component of $\mathcal{R}(\phi^t)$ contains $\mathcal{O}(x)$. Since ϕ^t is strongly gradient-like $\mathcal{R}(\phi^t)$ is totally disconnected. That is, the connected components of $\mathcal{R}(\phi^t)$ are singletons. Thus, $\mathcal{O}(x) = \{x\}$. So, $x \in \mathrm{Fix}(\phi^t)$. Therefore,

$$\mathcal{R}(\phi^t) = \mathrm{Fix}(\phi^t).$$

□

4.7. Complete Lyapunov Functions

The results of Section 4.6 show that the gradient flow ϕ^t of a typical Morse function f on a compact Riemannian manifold M has the following important properties:

(1) $f(\phi^t(x))$ is a strictly decreasing function of t for all $x \in M\setminus\mathrm{Fix}(\phi^t)$,

(2) $f|_{\mathrm{Fix}(\phi^t)}$ is injective, and

(3) $f(\mathrm{Fix}(\phi^t))$ is finite.

Our objective is to develop a continuous version of a Morse function for a flow on a metric space. To extend the properties of typical Morse functions to this new setting, replace the fixed point set with the chain recurrent set, replace each fixed point with a chain component, and replace the finiteness condition on the image of the set of fixed points with a nowhere density condition on the image of the chain recurrent set.

Definition 4.7.1. Let ϕ^t be a flow on a metric space X. A **complete Lyapunov function** for ϕ^t is a continuous function $\mathcal{L} : X \to \mathbb{R}$ such that

(1) $\mathcal{L}(\phi^t(x))$ is a strictly decreasing function of t for all $x \in X \setminus \mathcal{R}(\phi^t)$,

(2) for all $x, y \in \mathcal{R}(\phi^t)$ the points x and y are chain equivalent with respect to ϕ^t if and only if $\mathcal{L}(x) = \mathcal{L}(y)$, and

(3) $\mathcal{L}(\mathcal{R}(\phi^t))$ is nowhere dense.

The definition of a complete Lyapunov function is due to Conley [**11**].

Example 4.7.2. Every complete Lyapunov function is a weak Lyapunov function.

Examples 4.7.3 and 4.7.4 show that not every complete Lyapunov function is a strong Lyapunov function.

Example 4.7.3. A rational or irrational flow on the torus has one chain component — the entire torus. Therefore, the only complete Lyapunov functions for rational or irrational flows on the torus are constant functions.

Example 4.7.4. In Example 2.7.6 we saw that the arrested rotation has a single chain component, namely S^1. Therefore, the only complete Lyapunov functions for the arrested rotation are constant functions.

Example 4.7.5. Let ϕ^t be the decreasing segment flow on $[a, b]$. Example 2.7.7 showed that $\mathcal{R}(\phi^t) = \{a, b\}$. The chain components of this flow are $\{a\}$ and $\{b\}$. The identity function on $[a, b]$ is a complete Lyapunov function for this flow.

Example 4.7.6. Let ϕ^t be the arrested logistic rotation-dilation on the closed unit disk $\mathbb{D}$ in the complex plane. From Example 2.7.9 we know that $\mathcal{R}(\phi^t) = \partial\mathbb{D} \cup \{0\}$. The chain components of this flow are $\{0\}$ and $\partial\mathbb{D}$. The function $L : \mathbb{D} \to \mathbb{R}$ defined by $L(z) = 1 - |z|$ is a complete Lyapunov function for this flow.

Example 4.7.7. Let $\mathbb{T}^2$ be the torus represented by $\mathbb{R}^2/\mathbb{Z}^2$. The twist flow on the torus corresponds to the system of differential equations

$$\begin{aligned} \dot{x}_1 &= 1, \\ \dot{x}_2 &= \sin(2\pi x_2). \end{aligned}$$

The chain recurrent set of the twist flow consists of the segments $x_2 = 0$ and $x_2 = \frac{1}{2}$. We claim that a complete Lyapunov function for the twist flow is

$$\mathcal{L}(x_1, x_2) = \left(x_2 - \frac{1}{2}\right)^2.$$

If $(x_1(t), x_2(t))$ is a solution that is not chain recurrent, then

$$\begin{aligned} \frac{d}{dt}\mathcal{L}(x_1(t), x_2(t)) &= 2\left[x_2(t) - \frac{1}{2}\right]\dot{x}_2(t) \\ &= 2\left[x_2(t) - \frac{1}{2}\right]\sin[2\pi x_2(t)] \\ &< 0. \end{aligned}$$

Thus, $\mathcal{L}$ strictly decreases along orbits not in the chain recurrent set. Clearly, $\mathcal{L}$ is constant on each chain component and assumes distinct values on distinct chain components. The image of the chain recurrent set under $\mathcal{L}$ is $\{0, \frac{1}{2}\}$ which is finite, hence nowhere dense in $\mathbb{R}$.

4.8. The Fundamental Theorem of Dynamical Systems

The goal of this section is to prove the following extraordinary result due to Conley [**11**].

Theorem 4.8.1. The Fundamental Theorem of Dynamical Systems. *Every flow on a compact metric space has a complete Lyapunov function.*

The first step in proving the Fundamental Theorem is to demonstrate that the set of attracting-repelling pairs for a flow is at most countable (Proposition 4.8.6).

Lemma 4.8.2. *Let ϕ^t be a flow on a compact metric space X. Let A be the attracting set corresponding to a trapping region U with respect to ϕ^t. If $\mathcal{G}$ is an open subset of X such that $A \subseteq \mathcal{G}$, then there exists a positive real number T such that*

$$\bigcup_{t \geq T} \phi^t(U) \subseteq \mathcal{G}.$$

Proof. By means of contradiction assume that there does not exist a positive real number T such that $\bigcup_{t\geq T} \phi^t(U) \subseteq \mathcal{G}$. Thus, for each natural number k, there exist $t_k \geq k$ and $u_k \in U$ such that $\phi^{t_k}(u_k) \notin \mathcal{G}$. Since X is compact, the sequence $\phi^{t_k}(u_k)$ has a convergent subsequence. By reindexing if necessary, we may assume that $\phi^{t_k}(u_k)$ converges to some point x. Since $\mathcal{G}$ is open, the point x is not an element of $\mathcal{G}$. If $k > \tau \geq 0$, then $\phi^{t_k}(u_k) \in \phi^{\tau}(\mathrm{cl}(U))$ because U is a trapping region. Hence, $x \in \phi^{\tau}(\mathrm{cl}(U))$

for all $\tau \geq 0$. Since A is the attracting set corresponding to U, Proposition 3.4.15 implies that

$$A = \bigcap_{\tau \geq 0} \phi^{\tau}(\mathrm{cl}(U)) \subseteq \mathcal{G}.$$

Thus, $x \in A \subseteq \mathcal{G}$, contradicting the fact that $x \notin \mathcal{G}$. Therefore, there exists $T > 0$ such that

$$\bigcup_{t \geq T} \phi^t(U) \subseteq \mathcal{G}. \qquad \square$$

The following lemma is a version of Lemma 4.8.2 in backward time.

Lemma 4.8.3. *Let ϕ^t be a flow on a compact metric space X. Let A^* be the repelling set corresponding to a trapping region U with respect to ϕ^t. If $\mathcal{K}$ is a closed subset of X and $\mathcal{K} \cap A^* = \emptyset$, then there exists a positive real number T such that*

$$\bigcup_{t \geq T} \phi^t(\mathcal{K}) \subseteq U.$$

Proof. By Proposition 3.4.14 the set A^* is an attracting set for the trapping region $X \setminus U$ with respect to the flow ϕ^{-t}. Since A^* and $\mathcal{K}$ are disjoint, $X \setminus \mathcal{K}$ is an open subset of X such that $A^* \subseteq X \setminus \mathcal{K}$. By Lemma 4.8.2 there exists a positive real number T such that

$$\bigcup_{t \geq T} \phi^{-t}(X \setminus U) \subseteq X \setminus \mathcal{K}.$$

Thus,

$$X \setminus \bigcap_{t \geq T} \phi^{-t}(U) \subseteq X \setminus \mathcal{K},$$

so that

$$\mathcal{K} \subseteq \bigcap_{t \geq T} \phi^{-t}(U).$$

Therefore,

$$\bigcup_{t \geq T} \phi^t(\mathcal{K}) \subseteq U. \qquad \square$$

Together, Lemma 4.8.2 and Lemma 4.8.3 imply the following corollary.

Corollary 4.8.4. *Let ϕ^t be a flow on a compact metric space X. Let (A, A^*) be an attracting-repelling pair with respect to ϕ^t. If $\mathcal{K}$ is a closed subset of X such that $\mathcal{K} \cap A^* = \emptyset$, and $\mathcal{G}$ is an open subset of X such that $A \subseteq \mathcal{G}$, then there exists a positive real number T such that*

$$\bigcup_{t \geq T} \phi^t(\mathcal{K}) \subseteq \mathcal{G}.$$

Proof. By Lemma 4.8.2, there exists a positive real number T_1 such that

$$\bigcup_{t \geq T_1} \phi^t(U) \subseteq \mathcal{G}.$$

By Lemma 4.8.3, there exists a positive real number T_2 such that

$$\bigcup_{t \geq T_2} \phi^t(\mathcal{K}) \subseteq U.$$

Let $T = T_1 + T_2$. If $t \geq T$, then

$$\phi^t(\mathcal{K}) = \phi^{t-T_2}\left(\phi^{T_2}(\mathcal{K})\right) \subseteq \phi^{T_1}(U) \subseteq \mathcal{G}.$$

Therefore,

$$\bigcup_{t \geq T} \phi^t(\mathcal{K}) \subseteq \mathcal{G}.$$

□

Applying Corollary 4.8.4 to $\mathcal{K} = \mathrm{cl}(\mathcal{G})$ we obtain the following result.

Corollary 4.8.5. *Let ϕ^t be a flow on a compact metric space X. Let (A, A^*) be an attracting-repelling pair with respect to ϕ^t. If $\mathcal{G}$ is an open subset of X such that $A \subseteq \mathcal{G}$ and $cl(\mathcal{G}) \cap A^* = \emptyset$, then there exists a positive real number T such that*

$$\bigcup_{t \geq T} \phi^t(cl(\mathcal{G})) \subseteq \mathcal{G}.$$

Proposition 4.8.6. *If ϕ^t is a flow on a compact metric space X, then the set of attracting-repelling pairs for ϕ^t is countable.*

Proof. Because X is a metric space, there exists a countable basis $\mathcal{B}$ for the topology of X. Since A is an attracting set with respect to ϕ^t, there exists an open subset U of X such that

$$A = \bigcap_{t \geq 0} \phi^t(\mathrm{cl}(U)).$$

The compactness of X implies that the closed set A is compact. Thus, there exists a finite open covering $\{U_i\}_{i=1}^N \subseteq \mathcal{B}$ of A such that $U_i \subseteq U$ for $i = 1, \ldots, N$. Let

$$\mathcal{G} = \bigcup_{i=1}^N U_i.$$

Since $A \subseteq \mathcal{G}$ and $\mathcal{G}$ is open, Lemma 4.8.2 implies that there exists a positive real number T such that

$$\bigcup_{t \geq T} \phi^t(U) \subseteq \mathcal{G}.$$

So,

$$\bigcup_{t \geq T} \phi^t\left(\mathrm{cl}(U)\right) \subseteq \mathrm{cl}(\mathcal{G}).$$

Since $\mathcal{G} \subseteq U$,

$$A \subseteq \bigcap_{t \geq T} \phi^t(\mathrm{cl}(U)) \subseteq \bigcap_{t \geq 0} \phi^t(\mathrm{cl}(\mathcal{G})) \subseteq \bigcap_{t \geq 0} \phi^t(\mathrm{cl}(U)) = A.$$

Therefore,

$$A = \bigcap_{t \geq 0} \phi^t(\mathrm{cl}(\mathcal{G})).$$

Because $\mathcal{G}$ is a union of a finitely many open sets from the countable basis $\mathcal{B}$, distinct attracting sets correspond to distinct finite subsets of $\mathcal{B}$. Thus, there are countably many attracting sets. Similarly, there are countably many repelling sets. Therefore, the set of attracting-repelling pairs for ϕ^t is countable. □

Despite the fact that a flow on a compact space has at most countably many attracting-repelling pairs, Example 2.7.8 illustrates that such a flow may have uncountably many chain components.

The next step in proving the Fundamental Theorem is to show that for each attracting-repelling pair (A, A^*) there is a weak Lyapunov function L with respect to the flow such that

(1) $L\,|_A = 0$,
(2) $L\,|_{A^*} = 1$,
(3) $L\,(X \setminus (A \cup A^*)) \subseteq (0,1)$, and
(4) L strictly decreases along orbits not in $A \cup A^*$.

Proposition 4.8.7. *Let ϕ^t be a flow on a compact metric space (X, d). Let (A, A^*) be an attracting-repelling pair for ϕ^t. Define $V : X \to [0,1]$ by*

$$V(x) = \frac{d(x, A)}{d(x, A) + d(x, A^*)}.$$

Each of the following statements is true.

(i) $V\,|_A = 0$.
(ii) $V\,|_{A^*} = 1$.
(iii) $V\,(X \setminus (A \cup A^*)) \subseteq (0,1)$.
(iv) V *is continuous.*

Proof. The first three properties follow immediately from the definition of V, and the last follows from the continuity of the distance function. □

Define $V^* : X \to [0,1]$ by

$$V^*(x) = \sup\{V(\phi^t(x)) \mid t \geq 0\}.$$

The invariance of the sets A and A^* implies that $V^* \mid_A = 0$ and $V^* \mid_{A^*} = 1$. The function V^* is decreasing, though not necessarily strictly decreasing, along orbits in the complement of $A \cup A^*$. We will show that V^* is continuous.

Lemma 4.8.8. *Let ϕ^t be a flow on a compact metric space (X, d). Let (A, A^*) be an attracting-repelling pair for ϕ^t. Define $V : X \to [0, 1]$ by*

$$V(x) = \frac{d(x, A)}{d(x, A) + d(x, A^*)},$$

and define $V^* : X \to [0, 1]$ by

$$V^*(x) = \sup\{V(\phi^t(x)) \mid t \geq 0\}.$$

(i) *If $x \in X \setminus (A \cup A^*)$, then*

$$\lim_{t \to \infty} V\left(\phi^t(x)\right) = 0.$$

(ii) *V attains its supremum as a function of t for each $x \in X$. That is,*

$$V^*(x) = \max\{V\left(\phi^t(x)\right) \mid t \geq 0\}.$$

(iii) *$V^*(x) = 1$ if and only if $x \in A^*$.*

Proof. (i) Let $\epsilon > 0$. By Proposition 4.8.7 (i),(iv) the function V is continuous and $V|_A = 0$. So, for each $y \in A$ there exists $\delta_y > 0$ such that if $z \in X$ and $d(z, y) < \delta_y$, then $V(z) < \epsilon$. The collection $\{B(y, \delta_y) \mid y \in A\}$ is an open cover of A. The set A is compact because A is closed and X is compact. Thus, there exist $y_1, \ldots y_n \in A$ such that $\{B(y_i, \delta_{y_i}) \mid i = 1, \ldots, n\}$ is an open cover of A. Let $\delta = \min\{\delta_{y_1}, \ldots, \delta_{y_n}\}$.

Let $x \in X \setminus (A \cup A^*)$. By Proposition 2.3.6 (viii) we have

$$\lim_{t \to \infty} d(\phi^t(x), \omega(x)) = 0.$$

By Proposition 3.4.20 we know that $\omega(x) \subseteq A$. Hence,

$$\lim_{t \to \infty} d(\phi^t(x), A) = 0.$$

Consequently, there exists $T > 0$ such that $d(\phi^t(x), A) < \delta$ for all $t \geq T$. Thus, $V(\phi^t(x)) < \epsilon$ for all $t \geq T$. Therefore,

$$\lim_{t \to \infty} V(\phi^t(x)) = 0.$$

(ii) Let $x \in X$. If $x \in A \cup A^*$, then V is constant along the orbit of x so that V trivially attains its supremum. If $x \in X \setminus (A \cup A^*)$, then, by part (i), there exists $T \geq 0$ such that

$$V(\phi^t(x)) < \frac{1}{2} \sup\{V(\phi^t(x)) \,|\, t \geq 0\}$$

for all $t > T$. So,

$$\sup\{V(\phi^t(x)) \,|\, t \geq 0\} = \max\{V(\phi^t(x)) \mid 0 \leq t \leq T\}.$$

Thus, $V(\phi^t(x))$ attains its supremum as a function of t, and

$$V^*(x) = \max\{V(\phi^t(x)) \mid t \geq 0\}.$$

(iii) By part (ii), $V^*(x) = 1$ if and only if $V(\phi^\tau(x)) = 1$ for some $\tau \geq 0$. But, $V(\phi^\tau(x)) = 1$ if and only if $\phi^\tau(x) \in A^*$. Therefore, the invariance of A^* implies that $V^*(x) = 1$ if and only if $x \in A^*$. □

Lemma 4.8.9. *Let ϕ^t be a flow on a compact metric space (X, d). Let (A, A^*) be an attracting-repelling pair for ϕ^t. Define $V : X \to [0,1]$ by*

$$V(x) = \frac{d(x, A)}{d(x, A) + d(x, A^*)},$$

and define $V^* : X \to [0,1]$ by

$$V^*(x) = \sup\{V(\phi^t(x)) \mid t \geq 0\}.$$

The function V^* is continuous.

Proof. Since $V(x) \leq V^*(x) \leq 1$ for all $x \in X$, the continuity of V and the fact that $V|_{A^*} = 1$ imply that V^* is continuous on A^*.

Next, we show that V^* is continuous on A. Let $x \in A$, and let $\epsilon \in (0, \frac{1}{2})$. Define

$$\mathcal{G} = V^{-1}((-\infty, \epsilon)).$$

Since V is continuous, $\mathcal{G}$ is an open set. Because $A \subseteq \mathcal{G}$ and $\mathrm{cl}(\mathcal{G}) \cap A^* = \emptyset$, Corollary 4.8.5 implies that there exists a positive real number T such that

$$\bigcap_{s \geq T} \phi^s(\mathrm{cl}(\mathcal{G})) \subseteq \mathcal{G}.$$

Hence,

$$\bigcap_{t \geq 0} \phi^t\left(\phi^T(\mathcal{G})\right) \subseteq \mathcal{G}.$$

Since A is a closed invariant set contained in $\mathcal{G}$, $A \subseteq \phi^T(\mathcal{G})$. Thus, there exists $\delta > 0$ such that if $d(x, y) < \delta$, then $y \in \phi^T(\mathcal{G})$. Since $\phi^t(\phi^T(\mathcal{G})) \subseteq \mathcal{G}$ for all $t \geq 0$, if $d(x, y) < \delta$, then $\phi^t(y) \in \mathcal{G}$ for all $t \geq 0$. So, $V^*(y) < \epsilon$. By definition, $V^*(x) = 0$. Thus, if $d(x, y) < \delta$, then

$$|V^*(x) - V^*(y)| = V^*(y) < \epsilon.$$

Therefore, V^* is continuous on A.

Finally, we show that V^* is continuous on $X \setminus (A \cup A^*)$. Let $x \in X \setminus (A \cup A^*)$. Since A^* is closed and V is continuous, there exists $\eta > 0$ such that if $d(x, y) \leq 4\eta$, then $y \notin A^*$ and

$$V(y) \geq \tfrac{3}{4}V(x) > 0.$$

Let

$$\mathcal{G} = V^{-1}((-\infty, \tfrac{1}{2}V(x))).$$

Since $A \subseteq \mathcal{G}$ and $\overline{B}(x, \eta) \cap A^* = \emptyset$, Corollary 4.8.4 implies that there exists a positive real number T such that

$$\bigcap_{t \geq T} \phi^t(\overline{B}(x, \eta)) \subseteq \mathcal{G}.$$

So, if $y \in \overline{B}(x, \eta)$, then $V(y) \geq \frac{3}{4}V(x)$, and $V(\phi^t(y)) < \frac{1}{2}V(x)$ for all $t \geq T$. Therefore,

$$V^*(y) = \max\{V(\phi^t(y)) \mid 0 \leq t \leq T\}$$

for every $y \in \overline{B}(x, \eta)$. In particular, $V^*(x) = V(\phi^\tau(x))$ for some $\tau \in [0, T]$.

Let $\epsilon > 0$. By the continuity of V and the continuity of ϕ^τ, there exists $\delta \in (0, \eta)$ such that if $d(x, y) < \delta$, then

$$|V(\phi^\tau(y)) - V(\phi^\tau(x))| < \epsilon.$$

If $d(x, y) < \delta$, then

$$V^*(y) > V^*(x) - \epsilon.$$

We will show that if $d(x, y) < \delta$, then

$$V^*(y) < V^*(x) + \epsilon.$$

By means of contradiction assume that there exists a sequence y_k in $B(x, \delta)$ converging to $x \in X$ such that

$$V^*(y_k) \geq V^*(x) + \epsilon$$

for each $k = 1, 2, 3, \ldots$. There exists a sequence τ_k in $[0, T]$ such that

$$V^*(y_k) = V(\phi^{t_k}(y_k)) \geq V^*(x) + \epsilon$$

for each $k = 1, 2, 3, \ldots$. Extracting a subsequence if necessary, we may assume that τ_k converges to $t \in [0, T]$. Thus,

$$V(\phi^{\tau_k}(y_k)) \to V(\phi^t(x)) \geq V^*(x) + \epsilon$$

as $k \to \infty$, contradicting the definition of V^*. Hence,

$$V^*(y) < V^*(x) + \epsilon.$$

Therefore, if $d(x, y) < \delta$, then

$$|V^*(x) - V^*(y)| < \epsilon.$$

So, V^* is continuous on $X \setminus (A \cup A^*)$. □

To summarize, there is a function $V^* : X \to [0, 1]$ such that

(1) $V^*|_A = 0$,

(2) $V^*|_{A^*} = 1$,

(3) $V^*(X \setminus (A \cup A^*)) \subseteq (0, 1)$,

(4) V^* is decreasing along orbits in $X \setminus (A \cup A^*)$, and

(5) V^* is continuous.

Now we will use V^* to obtain a function $L : X \to [0,1]$ which strictly decreases along orbits in the complement of $A \cup A^*$. Let $f : \mathbb{R}^+ \cup \{0\} \to \mathbb{R}^+$ be a strictly decreasing function such that

$$\int_0^\infty f(s)\,ds = 1.$$

For example, $f(s) = e^{-s}$. Since V^* is continuous and $0 \leq V^*(x) \leq 1$ for all $x \in X$, the improper integral

$$\int_0^\infty f(s)V^*(\phi^s(x))\,ds$$

converges, and

$$0 \leq \int_0^\infty f(s)V^*(\phi^s(x))\,ds \leq 1$$

for all $x \in X$. Define $L : X \to [0,1]$ by

$$L(x) = \int_0^\infty f(s)V^*(\phi^s(x))\,ds.$$

Proposition 4.8.10. *Let ϕ^t be a flow on a compact metric space (X, d). Let (A, A^*) be an attracting-repelling pair for ϕ^t. Define $V : X \to [0,1]$ by*

$$V(x) = \frac{d(x,A)}{d(x,A) + d(x,A^*)},$$

and define $V^* : X \to [0,1]$ by

$$V^*(x) = \sup\{V(\phi^t(x)) \mid t \geq 0\}.$$

If $f : \mathbb{R}^+ \cup \{0\} \to \mathbb{R}^+$ is strictly decreasing and

$$\int_0^\infty f(s)\,ds = 1,$$

then the function $L : X \to [0,1]$ defined by

$$L(x) = \int_0^\infty f(s)V^*(\phi^s(x))\,ds$$

has the following properties.

(i) $L|_A = 0$,

(ii) $L|_{A^*} = 1$,

(iii) $L(X \setminus (A \cup A^*)) \subseteq (0,1)$,

(iv) *L is strictly decreasing along orbits in $X \setminus A \cup A^*$, and*

(v) *L is continuous.*

Proof. Properties (i) and (ii) follow from the corresponding properties of V^*. If $x \in X \setminus (A \cup A^*)$, then $V^*(\mathcal{O}(x)) \subseteq (0,1)$, so that the continuity of V^* implies property (iii). Let $x \in X \setminus (A \cup A^*)$. The function L decreases along the orbit of x because V^* decreases along the orbit of x. To see that L strictly decreases along the orbit of x we will prove that $L(\phi^t(x)) < L(x)$ for all $t > 0$.

By Lemma 4.8.8 (i),

$$\lim_{s\to\infty} V(\phi^s(\phi^t(x))) = \lim_{s\to\infty} V(\phi^{s+t}(x)) = 0$$

for all $t > 0$. Thus, for each $t > 0$ there exists $S > 0$ such that

$$V^*\left(\phi^{s+t}(x)\right) < V^*\left(\phi^s(x)\right)$$

for all $s > S$. Consequently,

$$L(\phi^t(x)) = \int_0^\infty f(s)V^*(\phi^{s+t}(x))\,ds < \int_0^\infty f(s)V^*(\phi^s(x))\,ds = L(x)$$

for all $t > 0$. Therefore, L strictly decreases along orbits in $X \setminus (A \cup A^*)$.

Finally, we show that L is continuous. Let $x \in X$, and let $\epsilon > 0$. Since

$$\int_0^\infty f(s)\,ds$$

converges, there exists $T > 0$ such that

$$\int_T^\infty f(s)\,ds < \frac{\epsilon}{4}.$$

By the continuity of ϕ^t and the continuity of V^*, there exists $\delta > 0$ such that if $d(x,y) < \delta$, then

$$|V^*(\phi^s(x)) - V^*(\phi^s(y))| < \frac{\epsilon}{2}$$

for all $s \in [0,T]$. If $d(x,y) < \delta$, then

$$\begin{aligned}
|L(x) - L(y)| &= \left| \int_0^\infty f(s)V^*(\phi^s(x))\,ds - \int_0^\infty f(s)V^*(\phi^s(y))\,ds \right| \\
&\leq \int_0^T f(s)\,|V^*(\phi^s(x)) - V^*(\phi^s(y))|\,ds \\
&\qquad + \int_T^\infty f(s)\,|V^*(\phi^s(x)) - V^*(\phi^s(y))|\,ds \\
&\leq \frac{\epsilon}{2}\int_0^T f(s)\,ds + \quad 2\int_T^\infty f(s)\,ds \\
&< \frac{\epsilon}{2} + \frac{\epsilon}{2} \\
&= \epsilon.
\end{aligned}$$

Therefore, L is a continuous function. $\square$

Now we prove the Fundamental Theorem of Dynamical Systems (Theorem 4.8.1) which establishes that every flow on a compact metric space has a complete Lyapunov function.

Proof. By Proposition 4.8.6 a flow on a compact metric space has at most countably many attracting-repelling pairs. For each attracting-repelling pair (A_j, A_j^*) there is a function L_j satisfying the conclusion of Proposition 4.8.10. For each $x \in X$ consider the series

$$\sum_{j=1}^{\infty} \frac{2}{3^j} L_j(x),$$

summing finitely terms if there are finitely many attracting-repelling pairs. Since $0 \leq L_j(x) \leq 1$ for all $x \in X$ and L_j is continuous for each j, the function $\mathcal{L}$ is a sum of continuous functions which converges uniformly by the Weierstrass M-test. Therefore, the function $\mathcal{L} : X \to [0,1]$ defined by

$$\mathcal{L}(x) = \sum_{j=1}^{\infty} \frac{2}{3^j} L_j(x)$$

is continuous.

Each function L_j has the property that

$$L_j\left(\phi^t(x)\right) \leq L_j(x)$$

for all $x \in X$ and all $t \geq 0$. Therefore,

$$\sum_{j=1}^{\infty} \frac{2}{3^j} L_j\left(\phi^t(x)\right) \leq \sum_{j=1}^{\infty} \frac{2}{3^j} L_j(x)$$

for all $x \in X$ and all $t \geq 0$. If $x \notin \mathcal{R}(\phi^t)$, then there exists a positive integer k such that $x \notin A_k \cup A_k^*$ by Theorem 3.4.27. So,

$$L_k\left(\phi^t(x)\right) < L_k(x)$$

for all $x \in X \setminus \mathcal{R}(\phi^t)$ and all $t > 0$. Since each L_j is a nonnegative function and $\mathcal{L}$ is a weighted sum of the functions L_j, we obtain

$$\mathcal{L}(\phi^t(x)) < \mathcal{L}(x)$$

for all $x \in X \setminus \mathcal{R}(\phi^t)$ and all $t > 0$.

Next, let $x, y \in \mathcal{R}(\phi^t)$. By Proposition 3.4.30 the points x and y are chain equivalent with respect to ϕ^t if and only if for each attracting-repelling pair (A, A^*) for ϕ^t, either $x, y \in A$ or $x, y \in A^*$. Consequently, x and y are chain equivalent if and only if $L_j(x) = L_j(y)$ for every index j. By the uniqueness of ternary expansions containing only zeros and twos, x and y are chain equivalent with respect to ϕ^t if and only if $\mathcal{L}(x) = \mathcal{L}(y)$.

Finally, if $x \in \mathcal{R}(\phi^t)$, then $L_j(x) \in \{0,1\}$ for each j. So, $\mathcal{L}(x)$ is an element of $[0,1]$ whose ternary expansion contains only zeros or twos. Thus, $\mathcal{L}(\mathcal{R}(\phi^t))$ is a subset of the middle-third Cantor set, which is a nowhere dense subset of $[0,1]$. Therefore, $\mathcal{L}$ is a complete Lyapunov function with respect to ϕ^t. □

To illustrate the construction of the complete Lyapunov function in the proof of the Fundamental Theorem we revisit the twist flow on the torus.

Example 4.8.11. Let $\mathbb{T}^2$ be the torus represented by $\mathbb{R}^2/\mathbb{Z}^2$. The twist flow on the torus corresponds to the differential equations

$$\begin{aligned} \dot{x}_1 &= 1, \\ \dot{x}_2 &= \sin(2\pi x_2). \end{aligned}$$

The chain recurrent set of the twist flow consists of the segments $x_2 = 0$ and $x_2 = \frac{1}{2}$. In this case the only nontrivial attracting-repelling pair is (A, A^*) where

$$A = \{(x,y) \in \mathbb{T}^2 \mid y = \frac{1}{2}\}$$

and

$$A^* = \{(x,y) \in \mathbb{T}^2 \mid y = 0, 1\}.$$

To construct a complete Lyapunov function for the twist flow, define $V : \mathbb{T}^2 \to \mathbb{R}$ by

$$\begin{aligned} V(x,y) &= \frac{d((x,y),A)}{d((x,y),A) + d((x,y),A^*)} \\ &= \frac{\left|y - \frac{1}{2}\right|}{\frac{1}{2}} \\ &= |2y-1|. \end{aligned}$$

The function V is already a complete Lyapunov function for the twist flow. However, we continue with the construction in the proof of Theorem 4.8.1 for illustrative purposes.

Define $V^* : \mathbb{T}^2 \to \mathbb{R}$ by

$$V * (x,y) = \sup\{V(\phi^t(x,y)) \mid t \geq 0\}.$$

Since $|2x_2(t) - 1|$ is a decreasing function of t, it follows that

$$V^*(x,y) = |2y-1|.$$

Let $f : \mathbb{R}^+ \cup \{0\} \to \mathbb{R}^+$ be a strictly decreasing, continuous function such that

$$\int_0^\infty f(s)\,ds = 1.$$

Define $\mathcal{L} : \mathbb{T}^2 \to \mathbb{R}$ by

$$\begin{aligned}\mathcal{L}(x,y) &= \int_0^\infty f(s)V^*(x_1(s),x_2(s))\,ds \\ &= \int_0^\infty f(s)\,|2x_2(s)-1|\,ds.\end{aligned}$$

The solution to the differential equation

$$\dot{x}_2 = \sin(2\pi x_2)$$

with $x_2(0) = y \neq \frac{1}{2}$ satisfies

$$\sin(2\pi x_2(t)) = \frac{2\tan(\pi y)e^{2\pi t}}{1+\tan^2(\pi y)e^{4\pi t}}.$$

Hence,

$$\dot{x}_2(t) = \frac{2\tan(\pi y)e^{2\pi t}}{1+\tan^2(\pi y)e^{4\pi t}}.$$

Let $F(s)$ be an antiderivative for $f(s)$. If $y < \frac{1}{2}$, then $x_2(s) < \frac{1}{2}$ for all $s \geq 0$. Integrating by parts we obtain

$$\begin{aligned}\mathcal{L}(x,y) &= \int_0^\infty f(s)\,(1-2x_2(s))\,ds \\ &= \lim_{t\to\infty}\Big[F(s)\,(1-2x_2(s))\Big]_0^t + \int_0^\infty F(s)\,(2\dot{x}_2(s))\,ds \\ &= \lim_{t\to\infty}\Big[F(s)\,(1-2x_2(s))\Big]_0^t + \int_0^\infty F(s)\left(\frac{4\tan(\pi y)e^{2\pi s}}{1+\tan^2(\pi y)e^{4\pi s}}\right)ds.\end{aligned}$$

If $f(s) = 2\pi e^{-2\pi s}$, then $F(s) = -e^{-2\pi s}$, and

$$\begin{aligned}L(x,y) &= 1-2y-\int_0^\infty \frac{4\tan(\pi y)}{1+\tan^2(\pi y)e^{4\pi s}}\,ds \\ &= 1-2y-\frac{\tan(\pi y)\ln(\csc^2(\pi y))}{\pi}.\end{aligned}$$

Exploiting symmetry about $y = \frac{1}{2}$,

$$\mathcal{L}(x,y) = \begin{cases} 1-2y-\frac{1}{\pi}\tan(\pi y)\ln(\csc^2(\pi y)) & \text{if } x < 1/2, \\ 0 & \text{if } x = 1/2, \\ 2y-1+\frac{1}{\pi}\tan(\pi y)\ln(\csc^2(\pi y)) & \text{if } x > 1/2. \end{cases}$$

If $f(s) = 4\pi e^{-4\pi s}$, then $F(s) = -e^{-4\pi s}$, and

$$\mathcal{L}(x,y) = \begin{cases} 1-2y\left(1-\tan^2(\pi y)\right)+\frac{2}{\pi}\tan(\pi x) & \text{if } x < 1/2, \\ 0 & \text{if } x = 1/2, \\ 2y-1\left(1-\tan^2(\pi y)\right)-\frac{2}{\pi}\tan(\pi x) & \text{if } x > 1/2. \end{cases}$$

Let ϕ^t be a flow on a compact metric space X. Proposition 3.3.5 showed that the chain equivalence relation $\sim$ on $\mathcal{R}(\phi^t)$ defined by

$$x \sim y \text{ if and only if } x \text{ and } y \text{ are chain equivalent}$$

is an equivalence relation. Consider the relation $\simeq$ on X defined by

$$x \simeq y \text{ if and only if } x = y \text{ or } x \sim y.$$

The relation $\simeq$ is an equivalence relation on X. See Exercise 18. Since X is compact, the collection $\widetilde{X} = X/\simeq$ of equivalence classes of $\simeq$ is a compact metric space. See Exercise 19. Let $p : X \to \widetilde{X}$ be the canonical projection. Define $\Phi : \mathbb{R} \times \widetilde{X} \to \widetilde{X}$ by

$$\Phi(t, \tilde{x}) = p(\phi^t(x))$$

such that $\tilde{x} = p(x)$. The function Φ is well-defined and is a flow on $\widetilde{X}$. See Exercise 20. Moreover,

$$\mathcal{R}(\Phi^t) = \mathrm{Fix}(\Phi^t) = p(\mathcal{R}(\phi^t)).$$

By the Fundamental Theorem of Dynamical Systems there exists a complete Lyapunov function $\mathcal{L}$ for ϕ^t. Since $\mathcal{L}$ is constant on $p^{-1}(\{\tilde{x}\})$ for each $\tilde{x} \in \widetilde{X}$, the function $\mathcal{L}$ induces a continuous function $\widetilde{\mathcal{L}} : \widetilde{X} \to \mathbb{R}$ such that

$$\widetilde{\mathcal{L}} \circ p = \mathcal{L}.$$

If $I : \mathbb{R} \to \mathbb{R}$ is the identity map, then the following diagram commutes.

$$\begin{array}{ccccc} X & \xrightarrow{\phi^t} & X & \xrightarrow{\mathcal{L}} & \mathbb{R} \\ p\downarrow & & \downarrow p & & \downarrow I \\ \widetilde{X} & \xrightarrow[\Phi^t]{} & \widetilde{X} & \xrightarrow[\widetilde{\mathcal{L}}]{} & \mathbb{R} \end{array}$$

The function $\widetilde{\mathcal{L}}$ is a strong Lyapunov function for Φ^t. See Exercise 21. Thus, Φ^t is gradient-like. Therefore, every flow on a compact metric space X induces a gradient-like flow on the quotient $\widetilde{X}$.

Example 4.8.12. Let ψ^t be the arrested logistic rotation-dilation on the closed unit disk $\mathbb{D}$. From Examples 2.7.9 and 3.3.13 we know that

$$\mathcal{R}(\psi^t) = \partial\mathbb{D} \cup \{0\},$$

and the chain components of ϕ^t are $\{0\}$ and $\partial\mathbb{D}$. As a consequence of the Fundamental Theorem the induced flow on $S^2 = \mathbb{D}/\simeq$ is gradient-like. In fact, the induced flow on S^2 is the gradient-like flow of Example 4.4.11. See Figure 4.15.

The Fundamental Theorem of Dynamical Systems implies that every strongly gradient-like flow on a compact metric space is gradient-like.

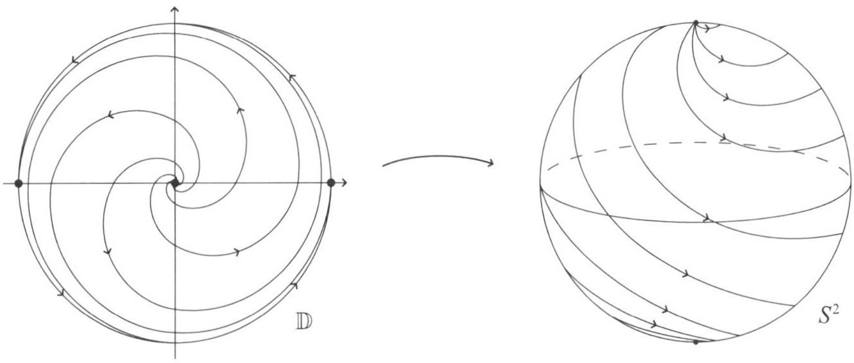

Figure 4.15. The gradient-like quotient of the arrested logistic rotation-dilation

Proposition 4.8.13. *Every strongly gradient-like flow on a compact metric space is gradient-like.*

Proof. Let ϕ^t be a strongly gradient-like flow on a compact metric space. According to the Fundamental Theorem of Dynamical Systems there exists a complete Lyapunov function $\mathcal{L}$ for ϕ^t. By Proposition 4.6.14,

$$\mathcal{R}(\phi^t) = \text{Fix}(\phi^t).$$

Consequently, $\mathcal{L}$ is a strong Lyapunov function with respect to ϕ^t. Therefore, ϕ^t is gradient-like. $\square$

4.9. Exercises

(1) For each of the following potential functions V sketch representative orbits of the flow corresponding to the differential equation

$$\dot{x} = -\nabla V(x)$$

on $\mathbb{R}^2$.

(a) $V(x, y) = -x^2 + y^2$,

(b) $V(x, y) = \dfrac{-2x^2 + x^4 + 2y^2}{4}$.

(c) $V(x, y) = \dfrac{2x^2 - x^4 + 2y^2}{4}$.

(d) $V(x, y) = \dfrac{3x^2 - 2x^3 + 3y^2}{6}$.

(2) Prove that every flow on $\mathbb{R}$ is a gradient flow.

(3) Show that the orthogonal projection of the vector

$$\begin{pmatrix} 0 \\ 0 \\ 1 \end{pmatrix}$$

to the tangent plane to S^2 at (x, y, z) is

$$\begin{pmatrix} -xz \\ -yz \\ x^2 + y^2 \end{pmatrix}.$$

(4) Consider the parametrization of the torus $\mathbb{T}^2$ in $\mathbb{R}^3$ given by

$$\begin{aligned} x(\alpha, \theta) &= (2 + \cos \alpha) \cos \theta, \\ y(\alpha, \theta) &= -\sin \alpha, \\ z(\alpha, \theta) &= (2 + \cos \alpha) \sin \theta \end{aligned}$$

where $\alpha, \theta \in [0, 2\pi]$. Show that the orthogonal projection of the vector

$$\begin{pmatrix} 0 \\ 0 \\ 1 \end{pmatrix}$$

to the tangent plane to $\mathbb{T}^2$ at the point with coordinates (α, θ) is

$$\begin{pmatrix} -\cos^2 \alpha \sin \theta \cos \theta \\ \sin \alpha \cos \alpha \sin \theta \\ \sin^2 \alpha \sin^2 \theta + \cos^2 \theta \end{pmatrix}.$$

(5) Find the chain recurrent set of the height flow on the sphere using Theorem 3.4.27.

(6) Prove that the flow of the system of differential equations

$$\begin{aligned} \dot{x} &= y, \\ \dot{y} &= x - x^3 \end{aligned}$$

is Hamiltonian, and sketch representative orbits of the flow.

(7) The **Poisson bracket** of the differentiable functions $f : \mathbb{R}^{2n} \to \mathbb{R}$ and $g : \mathbb{R}^{2n} \to \mathbb{R}$ is

$$[f, g] = \sum_{j=1}^{n} \frac{\partial f}{\partial q_j} \frac{\partial g}{\partial p_j} - \frac{\partial f}{\partial p_j} \frac{\partial g}{\partial q_j}.$$

(a) Let

$$J_n = \begin{pmatrix} 0_n & I_n \\ -I_n & 0_n \end{pmatrix}$$

where 0_n is the $n \times n$ zero matrix and I_n is the $n \times n$ identity matrix. Prove that

$$[f, g](q, p) = \nabla f_{(q,p)}^T J_n \nabla g_{(q,p)}.$$

(b) Prove that Hamilton's differential equations have the form

$$\begin{aligned} \dot{q}_j &= [q_j, H], \\ \dot{p}_j &= [p_j, H] \end{aligned}$$

for $j = 1, \ldots, n$.

(8) Let $A \in \mathbb{R}^{n \times n}$. Prove that if $M : \mathbb{R} \to \mathbb{R}^{n \times n}$ is a fundamental matrix solution of the linear differential equation $\dot{x} = Ax$, then there exists $t_0 \in \mathbb{R}$ such that

$$e^{At} = M(t)M(t_0)^{-1}$$

for all $t \in \mathbb{R}$.

(9) Use the Poincaré Recurrence Theorem to prove that $\mathbb{T}^2$ is a minimal set for an irrational flow on $\mathbb{T}^2$.

(10) Let ϕ^t be a gradient flow on a Riemannian manifold M with potential function $f : M \to \mathbb{R}$. Prove that if $p \in \mathrm{Fix}(\phi^t)$, then every eigenvalue of $D(\mathrm{grad}(f))_p$ is real.

(11) Prove that the height function $f : S^2 \to \mathbb{R}$ defined by $f(x, y, z) = z$ is a Morse function.

(12) Prove that the height function $f : \mathbb{T}^2 \to \mathbb{R}$ defined by $f(x, y, z) = z$ is a Morse function.

(13) Prove or disprove: The set of Morse functions on a compact manifold M which assume distinct values at distinct critical points is open in the set of Morse functions on M with respect to the topology of uniform convergence.

(14) Provide an example of a gradient flow which is not strongly gradient-like.

(15) Let ϕ^t be a gradient flow on a compact Riemannian manifold M.
   (a) Prove that there exists a complete Lyapunov function $\mathcal{L}$ with respect to ϕ^t.
   (b) Prove that if V is a potential function for ϕ^t, then $V + \mathcal{L}$ is a complete Lyapunov function with respect to ϕ^t.
   (c) Prove that $\mathcal{R}(\phi^t) = \mathrm{Fix}(\phi^t)$. Cf. Proposition 4.5.5.
   (d) Prove that $\alpha(x) \cup \omega(x) \subseteq \mathrm{Fix}(\phi^t)$ for all $x \in M$. Cf. Proposition 4.1.12.

(16) Let ϕ^t be a flow on a topological space X with finitely many chain components $C_1, \ldots, C_n$. A **filtration** of X with respect to ϕ^t is a collection $\{X_0, \ldots, X_n\}$ of subsets of X such that
  (i) $\emptyset = X_0 \subseteq X_0 \subseteq X_1 \subseteq \cdots \subseteq X_n = X$,
  (ii) X_j is a trapping region for ϕ^t for $j = 0, \ldots, n$,
  (iii) $C_j \subseteq \operatorname{int}(X_j \setminus X_{j-1})$ for $j = 1, \ldots, n$, and
  (iv) $C_j = \bigcap_{t \in \mathbb{R}} \phi^t(X_j \setminus X_{j-1})$.

Prove that if ϕ^t is a flow on a topological space X with finitely many chain components, then X has a filtration with respect to ϕ^t.

(17) Let ϕ^t be a flow on a topological space X. Let A be an attracting set for ϕ^t. The **basin** of A is the union of all open trapping regions U for ϕ^t such that $A_U = A$.

Let ϕ^t be the flow corresponding to the system of differential equations

$$\begin{aligned} \dot{x} &= y, \\ \dot{y} &= -y + x - x^3 \end{aligned}$$

on $\mathbb{R}^2$.
  (a) Find $\operatorname{Fix}(\phi^t)$.
  (b) Define $\mathcal{L} : \mathbb{R}^2 \to \mathbb{R}$ by

$$L(x, y) = \frac{1}{2}y^2 - \frac{1}{2}x^2 + \frac{1}{4}x^4.$$

  Prove that $\mathcal{L}$ is a complete Lyapunov function for ϕ^t.
  (c) Prove that $\mathcal{R}(\phi^t) = \operatorname{Fix}(\phi^t)$.
  (d) Find the attracting sets for ϕ^t.
  (e) Find the basin of each attracting set for ϕ^t.

(18) Let ϕ^t be a flow on a compact metric space X. Prove that the relation $\simeq$ on X defined by $x \simeq y$ if and only if $x = y$ or x is chain equivalent to y with respect to ϕ^t is an equivalence relation on X.

(19) Let ϕ^t be a flow on a compact metric space (X, d). Let $\simeq$ be the relation on X defined by $x \simeq y$ if and only if $x = y$ or x is chain equivalent to y with respect to ϕ^t. Let $\widetilde{X} = X/\simeq$. Define $\tilde{d} : \widetilde{X} \times \widetilde{X} \to \mathbb{R}$ by

$$\tilde{d}(\tilde{x}, \tilde{y}) = D(\tilde{x}, \tilde{y})$$

where $D : \mathcal{H} \times \mathcal{H} \to \mathbb{R}$ is the Hausdorff metric on the collection of nonempty closed bounded subsets of (X, d). Prove that $(\widetilde{X}, \tilde{d})$ is a compact metric space.

(20) Let ϕ^t be a flow on a compact metric space X. Let $\simeq$ be the relation on X defined by $x \simeq y$ if and only if $x = y$ or x is chain equivalent to y with respect to ϕ^t. Let $\widetilde{X} = X/\simeq$. Let $p : X \to \widetilde{X}$ be the canonical projection. Define $\Phi : \mathbb{R} \times \widetilde{X} \to \widetilde{X}$ by

$$\Phi(t, \tilde{x}) = p(\phi^t(x))$$

such that $\tilde{x} = p(x)$. Prove that the function Φ is well-defined, is a flow on $\widetilde{X}$, and

$$\mathcal{R}(\phi^t) = \mathrm{Fix}(\phi^t) = p(\mathcal{R}(\phi^t)).$$

(21) Let ϕ^t be a flow on a compact metric space (X, d). Let $\simeq$ be the relation on X defined by $x \simeq y$ if and only if $x = y$ or x is chain equivalent to y with respect to ϕ^t. Let $\widetilde{X} = X/\simeq$. Let $p : X \to \widetilde{X}$ be the canonical projection. Consider the flow $\Phi^t : \widetilde{X} \to \widetilde{X}$ induced by ϕ^t. Prove that there exists a strong Lyapunov function $\widetilde{\mathcal{L}} : \widetilde{X} \to \mathbb{R}$ for Φ^t such that

$$\widetilde{\mathcal{L}} \circ p = \mathcal{L}.$$

Afterword

Let us retrace the primary thread of *Recurrence and Topology*. If ϕ^t is a flow on a compact metric space X, then the phase space X of ϕ^t partitions into two invariant subsets: the chain recurrent set $\mathcal{R}(\phi^t)$ and its complement $\mathcal{G} = X \setminus \mathcal{R}(\phi^t)$. The chain recurrent set of ϕ^t is a nonempty closed invariant set containing the recurrent orbits of ϕ^t. Furthermore, the chain equivalence relation partitions the chain recurrent set into chain components, each of which is a nonempty closed invariant subset that is dynamically irreducible with respect to chain recurrence and is topologically irreducible with respect to connectedness. In this way the dynamics of the flow and the topology of its chain recurrent set are compatible.

By Conley's Fundamental Theorem of Dynamical Systems, the flow ϕ^t admits a complete Lyapunov function $\mathcal{L}$. That is, the function $\mathcal{L}$ is constant on each chain component, strictly decreases along orbits in $\mathcal{G}$, and assumes distinct values on distinct chain components. Because $\mathcal{L}$ is constant on each chain component, the flow ϕ^t restricted to its chain recurrent set resembles a Hamiltonian flow for which each chain component is analogous to a level set of a Hamiltonian function.

The fact that $\mathcal{L}$ strictly decreases along orbits in $\mathcal{G}$ and assumes distinct values on distinct chain components suggests that the flow ϕ^t restricted to $\mathcal{G}$ resembles a gradient flow arising from a generic Morse potential function in the sense that ϕ^t induces a gradient-like flow on the quotient of X obtained by identifying each chain component to a point.

An example containing many of the salient features of this development is the arrested logistic rotation-dilation corresponding to the system of differential equations on the closed unit disk in the complex plane given in

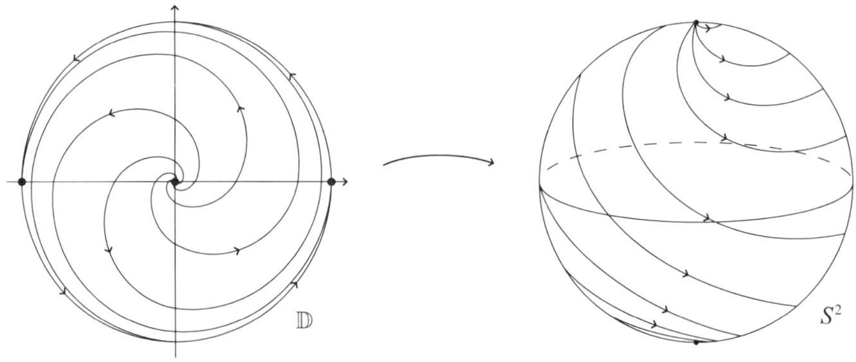

Figure 4.16. The arrested logistic rotation-dilation and its gradient-like quotient

polar coordinates by

$$\begin{aligned}\dot{r} &= r(1-r)(\sin^2\theta + 1 - r^2),\\ \dot{\theta} &= \sin^2\theta + 1 - r^2\end{aligned}$$

with $r \geq 0$. Figure 4.16 shows a phase portrait of this flow and its gradient-like quotient.

Conley introduced chain recurrence to study the coarse properties of phenomena which differential equations model only approximately. The epsilons in (ϵ, T)-chains capture the error inherent in such a system. In the spirit of the winding number and the Morse index, the Conley index is a stable algebraic topological invariant used to describe the coarse properties of a system. See Conley [**11**]. Franks and Richeson [**15**] develops the Conley index and its history.

Chain recurrence and the Fundamental Theorem have extensive applications to hyperbolic dynamical systems. The statements of many theorems in this area are very natural when expressed in terms of chain recurrence.

Let ϕ^t be a smooth flow on a compact manifold M. Roughly, an invariant set Λ for ϕ^t has a hyperbolic structure if at each point in Λ the tangent space to M splits into three subspaces: uniformly expanding and contracting subspaces and a one-dimensional central subspace with respect to the linearized flow. In the Shadowing Theorem, Bowen [**8**], [**9**] showed that for each (ϵ, T)-chain near enough to a hyperbolic invariant set there is an orbit that "shadows" the chain.

Flows with hyperbolic chain recurrent sets have particularly rich behavior. In this case, the Anosov Closing Lemma establishes that the chain recurrent set of ϕ^t is the closure of the periodic set of ϕ^t. This fact follows from the Shadowing Theorem and the Restriction Property of the chain recurrent set. The Spectral Decomposition Theorem implies that ϕ^t has finitely many chain components, often called basic sets. See Smale [**41**]. Together, the Spectral Decomposition Theorem and the Fundamental Theorem of Dynamical Systems imply the Ω-Stability Theorem: if ψ^t is a sufficiently small perturbation of ϕ^t, then the flows ϕ^t and ψ^t restricted to their respective chain recurrent sets are topologically conjugate. The Structural Stability Theorem obtains a global topological conjugacy between ϕ^t and ψ^t whenever the chain recurrent set of ϕ^t satisfies an additional transversality condition. The most general form of the Structural Stability Theorem is due to Robinson [**35**] and is a major achievement of the modern theory of dynamical systems. Robinson [**37**] extensively discusses chain recurrence in the presence of hyperbolicity. Franks [**14**] examines homology and hyperbolic dynamical systems using the Fundamental Theorem as a point of departure.

Interested readers may also pursue other directions. Robinson [**37**] and Katok and Hasselblatt [**26**] are comprehensive resources on dynamical systems. Each has voluminous references. Akin [**2**], Ellis [**13**] and Gottschalk and Hedlund [**16**] treat general topological dynamics in a more abstract setting. Abraham and Marsden [**1**] and Arnold [**3**] contain expositions of Hamiltonian dynamics, while Walters [**44**] studies more general dynamical systems preserving a measure such as volume.

John M. Alongi
Chicago, IL

Gail S. Nelson
Northfield, MN

Discrete Dynamical Systems

This appendix outlines some elementary ideas about discrete dynamical systems which are useful in Appendix B and when referring to time-one maps of flows.

Definition A.0.1. Let X be a set. A **map** (or **discrete dynamical system**) is a function $f : X \to X$.

Let$f$ be map on a set X. For each positive integer n let f^n denote the n-fold composition of f with itself, that is,

$$f^n = \underbrace{f \circ \cdots \circ f}_{n \text{ times}}.$$

If f is invertible with inverse f^{-1}, then for each positive integer n define

$$f^{-n} = (f^{-1})^n.$$

Denote the identity map on X by f^0.

Definition A.0.2. Let X be a set, $f : X \to X$, and $x \in X$. The **forward orbit** of x under f is the set

$$\mathcal{O}^+(x) = \{f^n(x) \mid n \in \mathbb{N}\}.$$

If f is invertible, then the **backward orbit** of x under f is the set

$$\mathcal{O}^-(x) = \{f^{-n}(x) \mid n \in \mathbb{N}\},$$

and the **orbit** of x under f is the set

$$\mathcal{O}(x) = \{f^n(x) \mid n \in \mathbb{Z}\}.$$

Example A.0.3. Define $f : \mathbb{R} \to \mathbb{R}$ by

$$f(x) = 2x.$$

Since f is invertible, the orbit of each real number x is

$$\mathcal{O}(x) = \{2^n x \mid n \in \mathbb{Z}\}.$$

As Example A.0.3 indicates, the orbit of a point under a map may not be a connected set.

Definition A.0.4. Let X be a set and $f : X \to X$. A subset A of X is called **positively invariant** with respect to f if $f(A) \subseteq A$. A subset A of X is called **negatively invariant** with respect to f if $f^{-1}(A) \subseteq A$. A subset A of X is called **invariant** with respect to f if A is positively invariant and negatively invariant.

In Definition A.0.4 the symbol $f^{-1}(A)$ denotes the inverse image of A under f. The definition does not demand that f be invertible.

The forward orbit of a point is a positively invariant set. The backward orbit of a point is a negatively invariant set, and the orbit of a point is an invariant set. The complement of a positively invariant set is negatively invariant, and the complement of a negatively invariant set is positively invariant. Consequently, complements of invariant sets are invariant.

Proposition A.0.5. *Let X be a first-countable topological space, and let $f : X \to X$ be a continuous function.*

(i) *If A is positively invariant with respect to f, then cl(A) is positively invariant with respect to f.*

(ii) *If A is negatively invariant with respect to f, then cl(A) is negatively invariant with respect to f.*

(iii) *If A is invariant with respect to f, then cl(A) is invariant with respect to f.*

Proof. (i) Since A is positively invariant, it suffices to prove that the set of limit points of A is positively invariant. Let x be a limit point of A. Since X is first-countable, there exists a sequence $\{x_i\}$ in A converging to x. Let n be a positive integer. Because A is positively invariant, $\{f^n(x_i)\}$ lies in A. Because X is first-countable and f is continuous, the sequence $\{f^n(x_i)\}$ converges to $f^n(x)$. Thus, $f^n(x)$ is a limit point of A. Consequently, $f^n(x) \in \mathrm{cl}A$. Therefore, $\mathrm{cl}(A)$ is positively invariant.

(ii) The proof is similar to (i).

(iii) If A is invariant, then A is positively invariant and negatively invariant. By (i) and (ii), $\mathrm{cl}(A)$ is positively invariant and negatively invariant. Therefore, $\mathrm{cl}(A)$ is invariant. $\square$

Circle Rotations

Example 2.3.10 and Lemma 2.7.19 employ facts from discrete dynamical systems about rotations of the circle $S^1 = \mathbb{R}/\mathbb{Z}$. We will show that every point is periodic with respect to a rational rotation of the circle and that the orbit of every point under an irrational rotation of the circle is dense in the circle.

Definition B.0.6. The **rotation of the circle** by a real number λ is the function $f_\lambda : S^1 \to S^1$ defined by

$$f_\lambda(x) = x + \lambda \bmod \mathbb{Z}.$$

For each real number λ the function $f_\lambda \colon S^1 \to S^1$ is continuous and has a continuous inverse f_λ^{-1} defined by

$$f_\lambda^{-1}(x) = x - \lambda \bmod \mathbb{Z}.$$

Notice that

$$f_\lambda^n(x) = x + n\lambda \bmod \mathbb{Z},$$

and f_λ^n is a homeomorphism for every integer n.

By an arc contained in S^1 we mean a connected subset of S^1. In preparation for the proof of Theorem B.0.8 we show that every open subset of S^1 contains an open arc of greatest length. Denote the length of an open arc γ contained in S^1 by $l(\gamma)$.

Lemma B.0.7. *If U is an open subset of S^1, then there exists an open arc γ contained in U such that the length of γ is at least the length of σ whenever σ is an open arc contained in U. If U is nonempty, then γ is nonempty.*

Proof. Let U be an open subset of S^1. There exists a countable collection $\Gamma = \{\gamma_i\}$ of disjoint open arcs whose union is U. If U is empty, then let γ be the empty set. If U is nonempty, then there exists $\alpha \in \Gamma$ such that $l(\alpha) > 0$. Define

$$s_n = \sum_{i=0}^{n} l(\gamma_i)$$

for each natural number n. Since Γ is a disjoint collection whose union is a subset of S^1, and S^1 has finite arc length, the sequence $\{s_n\}$ is bounded. Since $l(\gamma_i) \geq 0$ for each natural number i, the sequence $\{s_n\}$ is monotone. Thus, the sequence $\{s_n\}$ is convergent and is the sequence of partial sums of the series

$$\sum_{i=0}^{\infty} l(\gamma_i).$$

Thus, the series converges. Consequently,

$$\lim_{i\to\infty} l(\gamma_i) = 0.$$

Hence, there exists a natural number K such that

$$l(\gamma_k) < l(\alpha)$$

for all $k > K$. Let γ be an arc of greatest length in the finite collection $\{\gamma_0, \ldots, \gamma_K\}$. It follows that $l(\beta) \leq l(\gamma)$ for all $\beta \in \Gamma$. In particular, γ is nonempty because $l(\gamma) \geq l(\alpha) > 0$.

Let σ be an open arc contained in U. We claim that some element of Γ contains σ. Because Γ is a disjoint collection of open sets whose union is U, and $\sigma \subseteq U$ is open, $\{\sigma \cap \gamma_i\}$ is a disjoint collection of open sets whose union is σ. The connectedness of σ implies that this collection has exactly one nonempty element. Thus, some element $\beta \in \Gamma$ contains σ, from which we conclude that $l(\sigma) \leq l(\beta) \leq l(\gamma)$. Therefore, the length of γ is at least the length of σ whenever σ is an open arc contained in U. □

Under a rotation of the circle either every point is periodic or every orbit is dense in the circle.

Theorem B.0.8. *Let λ be a real number. Define $f_\lambda : S^1 \to S^1$ by*

$$f_\lambda(x) = x + \lambda \ mod\ \mathbb{Z}.$$

(i) *If λ is a rational number, then x is a periodic point with period q for every $x \in S^1$ where p and q are relatively prime integers such that $\lambda = \dfrac{p}{q}$ and $q > 0$.*

(ii) *If λ is an irrational number, then $\mathcal{O}^+(x)$, $\mathcal{O}^-(x)$, and $\mathcal{O}(x)$ are dense in S^1 for every $x \in S^1$.*

Proof. (i) If λ is a rational number, then $\lambda = \frac{p}{q}$ for some relatively prime integers p and q with $q > 0$. Then

$$f_\lambda^q(x) = x + q \cdot \frac{p}{q} \bmod \mathbb{Z} = x + p \bmod \mathbb{Z} = x$$

for every $x \in S^1$. Therefore, x is a periodic point with period q for every $x \in S^1$.

(ii) If λ is an irrational number, then we shall demonstrate that $\mathcal{O}^+(x)$ is dense in S^1 for every $x \in S^1$. By means of contradiction, assume that there exists $x \in S^1$ such that $\mathcal{O}^+(x)$ is not dense in S^1. Thus, $\mathrm{cl}(\mathcal{O}^+(x))$ is a proper subset of S^1. Since $S^1 \setminus \mathrm{cl}(\mathcal{O}^+(x))$ is nonempty and open, there is a nonempty open arc γ contained in $S^1 \setminus \mathrm{cl}(\mathcal{O}^+(x))$ of greatest length by Lemma B.0.7. We claim that $f_\lambda^m(\gamma)$ and $f_\lambda^n(\gamma)$ are disjoint for all distinct nonpositive integers m and n. By means of contradiction, assume that $f_\lambda^m(\gamma)$ and $f_\lambda^n(\gamma)$ are not disjoint for some distinct nonpositive integers m and n.

First, suppose that $f_\lambda^m(\gamma) = f_\lambda^n(\gamma)$. Since f_λ^m is a homeomorphism, $f_\lambda^m(\gamma)$ is a nonempty open arc. Let a be an endpoint of γ. Because rotations preserve orientation, $f_\lambda^m(a) = f_\lambda^n(a)$. Hence, $a + m\lambda = a + n\lambda \bmod \mathbb{Z}$. So, $(m - n)\lambda \in \mathbb{Z}$. Consequently, λ is rational — a contradiction. Thus, $f_\lambda^m(\gamma) \neq f_\lambda^n(\gamma)$.

Second, since rotations preserve arc lengths and $f_\lambda^m(\gamma) \neq f_\lambda^n(\gamma)$, the length of the open arc $f_\lambda^m(\gamma) \cup f_\lambda^n(\gamma)$ is greater than the length of γ. By Proposition A.0.5 the set $\mathrm{cl}(\mathcal{O}^+(x))$ is positively invariant. Thus, $S^1 \setminus \mathrm{cl}(\mathcal{O}^+(x))$ is negatively invariant and contains γ. Consequently, the set $S^1 \setminus \mathrm{cl}(\mathcal{O}^+(x))$ contains $f_\lambda^n(\gamma) \cup f_\lambda^m(\gamma)$. This contradicts the fact that γ has the greatest length of any open arc contained in $S^1 \setminus \mathrm{cl}(\mathcal{O}^+(x))$. Therefore, $f_\lambda^m(\gamma)$ and $f_\lambda^n(\gamma)$ are disjoint for all distinct nonpositive integers m and n.

The collection $\{f_\lambda^{-n}(\gamma) \mid n \in \mathbb{N}\}$ consists of an infinite number of disjoint arcs of equal length, contradicting the finiteness of the circle's arc length. Therefore, $\mathcal{O}^+(x)$ is dense in S^1 for every $x \in S^1$. A similar argument proves that $\mathcal{O}^-(x)$ is dense in S^1 for every $x \in S^1$.

Since $\mathcal{O}^+(x) \subseteq \mathcal{O}(x)$ for all $x \in S^1$, we conclude that $\mathcal{O}(x)$ is dense in S^1 for every $x \in S^1$. □

Theorem B.0.8 has many useful consequences. Beyond the theorem itself, we require only one.

Proposition B.0.9. *If α and δ are positive real numbers and λ is a positive irrational number, then there exist natural numbers J and K such that*

$$|\alpha + J\lambda - K| < \delta.$$

Proof. Suppose α and δ are positive real numbers and λ is a positive irrational number. Let $f_\lambda(x) = x + \lambda \bmod \mathbb{Z}$. By Proposition B.0.8 the forward orbit of α under f_λ is dense in S^1. Thus, there exists a natural number J such that $f_\lambda^J(\alpha)$ is an element of the ball of radius δ about $0 \in S^1 = \mathbb{R}/\mathbb{Z}$. Therefore, there exists a natural number K such that $|\alpha + J\lambda - K| < \delta$. □

Appendix C

The Hausdorff Metric

The goal of this appendix is to endow the collection of nonempty closed bounded subsets of a metric space with a metric topology. We use this topology in Proposition 2.7.21 to prove that if we restrict a flow to its chain recurrent set, then the chain recurrent set of the restricted flow coincides with the chain recurrent set of the unrestricted flow.

Let (X, d) be a metric space. For each $A \subseteq X$ and $\epsilon > 0$ define

$$U(A, \epsilon) = \{x \in X \mid d(x, a) < \epsilon \text{ for some } a \in A\}.$$

The set $U(A, \epsilon)$ is the collection of points in X which are within distance ϵ of some point in A. Notice that if $B \subseteq U(A, \delta)$ for some $\delta > 0$, then $B \subseteq U(A, \epsilon)$ for all $\epsilon \geq \delta$.

Let $\mathcal{H}$ denote the collection of nonempty closed bounded subsets of X. Define $D : \mathcal{H} \times \mathcal{H} \to \mathbb{R}$ by

$$D(A, B) = \inf\{\epsilon > 0 \mid A \subseteq U(B, \epsilon) \text{ and } B \subseteq U(A, \epsilon)\}.$$

The number $D(A, B)$ is the least real number with the property that $A \subseteq U(B, \epsilon)$ and $B \subseteq U(A, \epsilon)$ for all $\epsilon > D(A, B)$.

The following proposition shows that D is a metric on $\mathcal{H}$.

Proposition C.0.10. *If $\mathcal{H}$ is the collection of nonempty closed bounded subsets of a metric space (X, d), and $D : \mathcal{H} \times \mathcal{H} \to \mathbb{R}$ is defined by*

$$D(A, B) = \inf\{\epsilon > 0 \mid A \subseteq U(B, \epsilon) \text{ and } B \subseteq U(A, \epsilon)\},$$

where $U(A, \epsilon) = \{x \in X \mid d(x, a) < \epsilon$ for some $a \in A\}$, then $(\mathcal{H}, D)$ is a metric space.

Proof. Let $(A, B) \in \mathcal{H} \times \mathcal{H}$. Since $D(A, B)$ is the infimum of a set of positive real numbers, $D(A, B) \geq 0$. Furthermore, $D(A, B) = 0$ if and only if there exists a sequence $\{\epsilon_n\}$ of positive real numbers converging to 0 such that $A \subseteq U(B, \epsilon_n)$ and $B \subseteq U(A, \epsilon_n)$ for every natural number n. As A and B are closed sets, this is true if and only if $A \subseteq B$ and $B \subseteq A$. Thus, $D(A, B) \geq 0$ for all $(A, B) \in \mathcal{H} \times \mathcal{H}$, with equality if and only if $A = B$.

From the definition of D we have $D(A, B) = D(B, A)$ for all $(A, B) \in \mathcal{H} \times \mathcal{H}$.

It remains to establish the triangle inequality. Let $A, B, C \in \mathcal{H}$. Suppose that $\alpha > D(A, B)$ and $\beta > D(B, C)$, so that $A \subseteq U(B, \alpha), B \subseteq U(A, \alpha), B \subseteq U(C, \beta)$ and $C \subseteq U(B, \beta)$. Since d satisfies the triangle inequality,

$$A \subseteq U(B, \alpha) \subseteq U(C, \alpha + \beta),$$

and

$$C \subseteq U(B, \beta) \subseteq U(A, \alpha + \beta).$$

Hence,

$$\alpha + \beta \geq D(A, C).$$

By taking infima over all $\alpha > D(A, B)$ and all $\beta > D(B, C)$, it follows that

$$D(A, B) + D(B, C) \geq D(A, C)$$

for all $A, B, C \in \mathcal{H}$.

Therefore, D is a metric on $\mathcal{H}$. □

Definition C.0.11. If $\mathcal{H}$ is the collection of nonempty closed bounded subsets of a metric space (X, d), then the **Hausdorff metric** on $\mathcal{H}$ is the function $D : \mathcal{H} \times \mathcal{H} \to \mathbb{R}$ defined by

$$D(A, B) = \inf \{\epsilon > 0 \,|\, A \subseteq U(B, \epsilon) \text{ and } B \subseteq U(A, \epsilon)\},$$

where $U(A, \epsilon) = \{x \in X \,|\, d(x, a) < \epsilon \text{ for some } a \in A\}$.

Endowed with the topology induced by the Hausdorff metric, the collection of nonempty closed bounded subsets of a metric space X inherits certain properties from X. Among these inherited properties are total boundedness, completeness and compactness.

Definition C.0.12. A metric space X is called **totally bounded** if for every $\epsilon > 0$ there exists a finite covering of X by ϵ-balls.

Proposition C.0.13. *If D is the Hausdorff metric on the collection $\mathcal{H}$ of nonempty closed bounded subsets of a totally bounded metric space (X, d), then $(\mathcal{H}, D)$ is totally bounded.*

Proof. If X is the empty set, then $\mathcal{H}$ is the empty set, which is totally bounded.

If X is nonempty, then let $\epsilon > 0$. We must show that there exists a finite covering of $\mathcal{H}$ by ϵ-balls. Let $\delta \in (0, \epsilon)$. For each $x \in X$ denote the δ-ball around x in the metric d by $B_d(x, \delta)$. Since X is totally bounded, there exists a finite nonempty set $S \subseteq X$ such that $\{B_d(x, \delta) \mid x \in S\}$ covers X. Since S is finite and nonempty, $S \in \mathcal{H}$. Let $\mathcal{A}$ be the collection of nonempty subsets of S. The collection $\mathcal{A}$ is finite because S is finite. Moreover, each element of $\mathcal{A}$ is finite and nonempty, so that $\mathcal{A} \subseteq \mathcal{H}$. For each $A \in \mathcal{H}$ denote the ϵ-ball around A in the Hausdorff metric by $B_D(A, \epsilon)$. We shall prove that the finite collection $\{B_D(A, \epsilon) \mid A \in \mathcal{A}\}$ covers $\mathcal{H}$.

Let $C \in \mathcal{H}$, and define $A = S \cap U(C, \delta)$. The set A is nonempty because S is nonempty, and every point in C is within distance δ of some point in S. Thus, $A \in \mathcal{A}$, and $A \subseteq U(C, \delta)$.

Now we show that $C \subseteq U(A, \delta)$. Let $y \in C$. There exists $x \in S \cap B_d(y, \delta) \subseteq S \cap U(C, \delta) = A$. So, $y \in B_d(x, \delta) \subseteq U(A, \delta)$. Thus, $C \subseteq U(A, \delta)$.

By the definition of the Hausdorff metric, $D(A, C) \leq \delta < \epsilon$, so that $C \in B_D(A, \epsilon)$. Since C was an arbitrary element of $\mathcal{H}$, the finite collection $\{B_D(A, \epsilon) \mid A \in \mathcal{A}\}$ covers $\mathcal{H}$. Therefore, $(\mathcal{H}, D)$ is totally bounded. □

Proposition C.0.14. *If D is the Hausdorff metric on the collection $\mathcal{H}$ of nonempty closed bounded subsets of a complete metric space (X, d), then the metric space $(\mathcal{H}, D)$ is complete.*

Proof. Let $\{A_n\}$ be a Cauchy sequence in $(\mathcal{H}, D)$. By extracting a subsequence if necessary, we may assume that $D(A_n, A_{n+1}) < 1/2^n$ for all natural numbers n. Define

$$A = \left\{ a \in X \mid a = \lim_{n \to \infty} x_n \text{ with } x_n \in A_n \text{ and } d(x_n, x_{n+1}) < \frac{1}{2^n} \ \forall n \in \mathbb{N} \right\}.$$

We will prove that $\{A_n\}$ converges to $\mathrm{cl}(A)$ and that $\mathrm{cl}(A) \in \mathcal{H}$.

Since $D(A_n, A_{n+1}) < 1/2^n$ for all natural numbers n, there exists a sequence $\{x_n\}$ in X with $x_n \in A_n$ and $d(x_n, x_{n+1}) < 1/2^n$ for all natural numbers n. Thus, $\{x_n\}$ is a Cauchy sequence in X. The completeness of X guarantees that $\{x_n\}$ converges to some $a \in X$. Consequently, A is nonempty.

We will prove simultaneously that A is bounded and that $\{A_n\}$ converges to $\mathrm{cl}(A)$. Let $\epsilon > 0$. There exists a natural number N such that $1/2^n < \epsilon$ for all $n > N$. Let $n > N + 1$, and define $\delta = \epsilon - 1/2^{n-1}$. First, we will show that $\mathrm{cl}(A) \subseteq U(A_n, \epsilon)$.

If $z \in \mathrm{cl}(A)$, then there exists $a \in A$ such that $d(a, z) < \delta$ and there exists a sequence $\{x_i\}$ in X which converges to a. By the triangle inequality,

$$d(x_n, z) \leq \left[\sum_{i=n}^{m-1} d(x_i, x_{i+1})\right] + d(x_m, a) + d(a, z)$$

for all $m > n$. Taking the limit as $m \to \infty$,

$$d(x_n, z) < \delta + \sum_{i=n}^{\infty} \frac{1}{2^i} = \delta + \frac{1}{2^{n-1}} = \epsilon.$$

Thus, $z \in U(A_n, \epsilon)$. Consequently, $\mathrm{cl}(A) \subseteq U(A_n, \epsilon)$.

Since $A_n \in \mathcal{H}$, the set A_n is bounded. So, $U(A_n, \epsilon)$ is bounded. Because $\mathrm{cl}(A) \subseteq U(A_n, \epsilon)$, the set $\mathrm{cl}(A)$ is bounded. Therefore, $\mathrm{cl}(A) \in \mathcal{H}$.

Second, we will show that $A_n \subseteq U(\mathrm{cl}(A), \epsilon)$. Let $z \in A_n$. Since $D(A_i, A_{i+1}) < 1/2^i$ there exists a sequence $\{x_i\}$ in X such that $x_n = z$, $x_i \in A_i$, and $d(x_i, x_{i+1}) < 1/2^i$ for all natural numbers i. Thus, $\{x_i\}$ is a Cauchy sequence. Since X is complete, $\{x_i\}$ converges to some a. By definition, $a \in A$. By the triangle inequality,

$$d(z, a) \leq \left[\sum_{i=n}^{m-1} d(x_i, x_{i+1})\right] + d(x_m, a)$$

for all $m > n$. Taking the limit as $m \to \infty$,

$$d(z, a) \leq \sum_{i=n}^{\infty} \frac{1}{2^i} = \frac{1}{2^{n-1}} < \epsilon.$$

Thus $A_n \subseteq U(\mathrm{cl}(A), \epsilon)$.

Finally, since $\mathrm{cl}(A) \subseteq U(A_n, \epsilon)$ and $A_n \subseteq U(\mathrm{cl}(A), \epsilon)$,

$$D(\mathrm{cl}(A), A_n) < \epsilon.$$

Consequently, $\{A_n\}$ converges to $\mathrm{cl}(A) \in \mathcal{H}$. Therefore, the metric space $(\mathcal{H}, D)$ is complete. □

Proposition C.0.15. *If D is the Hausdorff metric on the collection $\mathcal{H}$ of nonempty closed bounded subsets of a compact metric space (X, d), then $(\mathcal{H}, D)$ is compact.*

Proof. A metric space is compact if and only if it is totally bounded and complete. Therefore, if (X, d) is compact, then $(\mathcal{H}, D)$ is compact by Propositions C.0.13 and C.0.14. □

Bibliography

[1] Ralph Abraham and Jerrold E. Marsden. *Foundations of Mechanics*. Benjamin/Cummings, Don Mills, Ontario, 1978.

[2] Ethan Akin. *The General Topology of Dynamical Systems*. American Mathematical Society, Providence, 1993.

[3] Vladimir I. Arnold. *Mathematical Methods of Classical Mechanics*. Springer-Verlag, New York, second edition, 1989.

[4] Vladimir I. Arnold. *Ordinary Differential Equations*. Springer-Verlag, New York, 1992.

[5] Ivar Bendixson. Sur les courbes définies par des équations différentielles. *Acta Mathematica*, 24:1–88, 1901.

[6] George Birkhoff. Quelques théorèmes sur le mouvement des systèmes dynamiques. *Bulletin de la Société Mathématique de France*, 40:305–323, 1912.

[7] George Birkhoff. *Dynamical Systems*. American Mathematical Society, Providence, 1927.

[8] Rufus Bowen. *Equilibrium States and the Ergodic Theory of Anosov Diffeomorphisms*. Springer-Verlag, New York, 1975.

[9] Rufus Bowen. ω-Limit sets for axiom A diffeomorphisms. *Journal of Differential Equations*, 18:333–339, 1975.

[10] William E. Boyce and Richard C. DiPrima. *Elementary Differential Equations*. Wiley, Hoboken, NJ, eighth edition, 2005.

[11] Charles Conley. *Isolated Invariant Sets and the Morse Index*. American Mathematical Society, Providence, 1978.

[12] Arnaud Denjoy. Sur les curbes définies par les équations differentielles à la surface du tore. *Journal de Mathématiques Pures et Appliquées*, 11:333–375, 1932.

[13] Robert Ellis. *Lectures on Topological Dynamics*. W.A. Benjamin, New York, 1969.

[14] John Franks. *Homology and Dynamical Systems*. American Mathematical Society, Providence, 1982.

[15] John Franks and David S. Richeson. Shift equivalence and the Conley index. *Transactions of the American Mathematical Society*, 352:3305–3322, 2000.

[16] W.H. Gottschalk and G.A. Hedlund. *Topological Dynamics*. American Mathematical Society, Providence, 1955.

[17] Victor Guillemin and Alan Pollack. *Differential Topology*. Prentice-Hall, Englewood Cliffs, 1974.

[18] Jack Hale. *Ordinary Differential Equations*. Wiley, New York, 1969.

[19] Paul R. Halmos. *Naive Set Theory*. Springer-Verlag, New York, 1974.

[20] Noel J. Hicks. *Notes on Differential Geometry*. Van Nostrand, Princeton, 1965.

[21] Michael Hurley. Chain recurrence and attraction in noncompact spaces. *Ergodic Theory and Dynamical Systems*, 11:709–729, 1991.

[22] Michael Hurley. Chain recurrence and attraction in noncompact spaces II. *Proceedings of the American Mathematical Society*, 115:1139–1148, 1992.

[23] Michael Hurley. Chain recurrence, semiflows, and gradients. *Journal of Dynamics and Differential Equations*, 7:437–456, 1995.

[24] Wilfred Kaplan. Regular curve families filling the plane, I. *Duke Mathematical Journal*, 7:154–185, 1940.

[25] Irving Kaplansky. *An Introduction to Differential Algebra*. Hermann, Paris, 1957.

[26] Anatole Katok and Boris Hasselblatt. *Introduction to the Modern Theory of Dynamical Systems*. Cambridge University Press, New York, 1995.

[27] Hellmuth Kneser. Reguläre kurvenscharen auf den ringflächen. *Mathematische Annalen*, 91:135–154, 1924.

[28] Serge Lang. *Real and Functional Analysis*. Springer-Verlag, New York, third edition, 1993.

[29] Lawrence Markus. Global structure of ordinary differential equations on the plane. *Transactions of the American Mathematical Society*, 76:127–148, 1954.

[30] John W. Milnor. *Morse Theory*. Princeton University Press, Princeton, 1963.

[31] John W. Milnor. *Topology from the Differentiable Viewpoint*. The University of Virginia Press, Charlottesville, 1969.

[32] James R. Munkres. *Topology*. Prentice-Hall, Upper Saddle River, 2000.

[33] V. Nemytskii and V.V. Stepanov. *Qualitative Theory of Differential Equations*. Dover, New York, 1989.

[34] Henri Poincaré. *Les méthodes nouvelles de la mécanique céleste Vols. 1–3*. Gautier-Villars, Paris, 1892.

[35] Clark Robinson. Structural stability of C^1 flows. *Lecture Notes in Mathematics*, 468:262–277, 1975.

[36] Clark Robinson. Stability theorems and hyperbolicity in dynamical systems. *Rocky Mountain Journal of Mathematics*, 7:425–437, 1977.

[37] Clark Robinson. *Dynamical Systems: Stability, Symbolic Dynamics, and Chaos*. CRC Press, Boca Raton, 1999.

[38] Walter Rudin. *Principles of Mathematical Analysis*. McGraw Hill, San Francisco, third edition, 1976.

[39] Arthur J. Schwartz. A generalization of the Poincaré-Bendixson theorem to closed two-dimensional manifolds. *American Journal of Mathematics*, 85:453–458, 1963.

[40] George Sell. *Topological Dynamics and Ordinary Differential Equations.* Von Nostrand-Reinhold, Princeton, 1971.

[41] Stephen Smale. Differentiable dynamical systems. *Bulletin of the American Mathematical Society*, 73:747–817, 1967.

[42] Michael Spivak. *Calculus on Manifolds: A Modern Approach to Classical Theorems of Advanced Calculus.* Addison-Wesley, Menlo Park, 1965.

[43] Michael Spivak. *A Comprehensive Introduction to Differential Geometry: Volume One.* Publish or Perish, Houston, 1999.

[44] Peter Walters. *An Introduction to Ergodic Theory.* Springer-Verlag, New York, 1982.

[45] Hassler Whitney. Regular families of curves. *Mathematische Annalen*, 34:244–270, 1933.

Index